Science and Technology Centers

Science and Technology Centers

Victor J. Danilov

The MIT Press
Cambridge, Massachusetts
London, England

This book was set in Linotron 202 Univers by DEKR Corporation and printed and bound by Murray Printing Co. in the United States of America.

Library of Congress Cataloging in Publication Data

Danilov, Victor J.
  Science and technology centers.

  Bibliography: p.
  Includes index.
  1. Science museums. 2. Industrial museums.
I. Title.
Q105.A1D36 1982   069'.95   82-8946
ISBN 0-262-04068-9   AACR2

To my lovely wife, Toni, and the more than 100 million people who have enjoyed Chicago's Museum of Science and Industry over the last half-century.

Contents

Preface

Science and technology centers are relative newcomers to the museum field. They have emerged primarily in the United States within the last fifty years. During this short period, they have become among the most popular and effective museums in the world.

These unusual institutions differ from traditional museums in a number of respects. They are concerned with furthering public understanding and appreciation of the physical and life sciences, engineering, technology, industry, and health and seek to accomplish this goal by making museums both enlightening and entertaining. They are best known for their contemporary rather than historic perspective and their reliance on participatory exhibit techniques rather than objects of intrinsic value.

Some traditionalists have questioned whether such centers really are "museums," since many do not have collections of artifacts or follow usually accepted museum practices. These critics see science and technology centers as science education centers, exhibition halls, and/or "scientific Disneylands."

Despite such professional skepticism, it can be said that science and technology centers have grown rapidly and filled the public need for inviting museums that explain scientific principles, technological applications, and, increasingly, social implications of scientific and technological developments in an understandable, interesting, and meaningful way.

This book traces the evolution of the science and technology center movement, describes the characteristics and operations of many such museums, and provides helpful information for those considering starting or operating science and technology centers.

Science and technology centers are appearing in many large urban areas and even smaller cities. They take numerous forms, each reflecting the needs, resources, and desires of the areas being served. There is no pattern for starting a science and technology center nor a formula for success. What has been accomplished can be attributed to the imagination, hard work, and support of the people who have made science and technology centers among the most exciting, popular, and useful cultural institutions in a rapidly changing world.

I

Introduction to Science Centers

1

What Is a Science and Technology Center?

The use of the term *science and technology centers* is relatively recent, but science and technology centers have been evolving since the turn of the century as science museums, technical museums, industrial museums, health museums, and science and technology museums. Science and technology centers—sometimes merely called *science centers*—differ from traditional museums in several respects. For example, they are not object oriented, and they usually do not have curators, conduct research, or publish learned papers. They are basically contemporary, participatory, informal educational instruments rather than historic, "hands-off" respositories of artifacts. Unlike many museums that are quiet and elitist, science and technology centers are lively and populist. They seek to further public understanding of science and technology in an enlightening and entertaining manner and do not require any special interest or background to be understood or appreciated by the average person.

Lee Kimche, former director of the Institute of Museum Services, has stated that "Science centers provide a whole new field of self-motivating experiences in learning, through environmental exhibits that appeal to the senses, emotions, and intellect. They are among the most rapidly developing institutions of learning in contemporary society."[1] The "hands-on" approach has become so popular that the Educational Facilities Laboratories published a report in 1975 that said

In the past, museums earned the reputation of being the somewhat stodgy keepers of our culture's most prized objects. Docents led school children through the museum's halls with occasional admonishments to be quiet and orderly. On weekends, families arrived, the parents pulling kids along to expose them to a genuine cultural experience.

But today there exists a small but growing number of museums that are experience-oriented rather than object-oriented. These institutions are primarily concerned with bringing the visitor and the museum's resources together in such a way that learning can occur.

A concept basic to this approach is that the visitor can learn the most through an experience in which he is an active participant. Such participation may take the form of touching artifacts or live animals, trying out an experiment, measuring your own pulse rate, going on a fossil dig, or playing a game with a computer. In short, it's a hands-on, get-involved approach, rather than a hands-off, just-look approach.[2]

An article in *The Chronicle of Higher Education* called the science center "a burgeoning new concept in the museum world" that "beckons both children and adults to the world of science through do-it-yourself experiments and a fascinating array of biological, physical, and chemical phenomena."[3]

In a way, science and technology centers are a response to the changing attitudes, needs, and expectations of the visiting public. As Kenneth Hudson, who has written extensively on museums, pointed out in his book on museums for the 1980s,

During the past 25 years especially, the museum-going public has changed a great deal, and it is still changing. Its range of interests has widened, it is far less reverent and respectful in its attitudes, it expects to find electronic and other modern technical facilities adequately used, it distinguishes less and less between a museum and an exhibition, it considers the intellect to be no more prestigious or respectable than the emotions, and it sees no reason to pay attention to the subject-division and specialisms which are so dear to academics. This is a reflection of a fundamental change in thought and behaviour throughout the world, and in all fields of activity....[4]

The tremendous growth and contribution of science centers have occurred as the world searches for better solutions to cultural interpretation, public education, equal opportunity, social mobility, population explosion, food production, improved health, human rights, scientific and technological advances, and other issues of the twentieth century.

Science and technology centers are concerned primarily with the natural and applied sciences and usually cover such fields as physics, chemistry, biology, geology, astronomy, mathematics, engineering, and medicine. However, they frequently include other fields of local interest. In general, these unconventional museums seek to explain scientific principles, technological applications, and social implications of these principles and applications. "Their reasons for being are to explain science and scientific advances to the layman, to improve science education, and to interest young people in careers in science and engineering," says Alvin Schwartz.[5]

No two science and technology centers are alike. Some emphasize the physical sciences; others are largely concerned with the life sciences. Sometimes a science center includes natural history and occasionally even art or history. A science center also can be a children's museum or nature center.

Once considered the "ugly ducklings" of the museum world, science and technology centers increasingly are being recognized as pacesetters in new museum techniques. As stated by John Whitman in a *Museum News* article, "The key word in most definitions and descriptions of museums is 'object.' Generally, the objects in art, history, and other museums are the beautiful, the priceless, the unique. By contrast, the objects in science museums are nuts and bolts: machines, laboratory test tubes, flashing strobe lights and beaming lasers, buttons, buzzers and bells, coils and springs, levers, gears and pulleys—none unique nor especially valuable by esthetic standards, and most easily duplicated in as many reproductions as parts allow."[6]

These "nuts and bolts" museums usually have some historic objects, but some of the newer science centers have no artifacts and consist entirely of constructed exhibits for educational purposes. The principal reason for this difference is that science centers often are dealing with the intangible—the concepts and processes of science and technology.

Michael Templeton, executive director of the Oregon Museum of Science and Industry in Portland and former head of the Association of Science-Technology Centers, has sought to make this distinction: "Museums communicate our cultural heritage through the use of objects, experiences, and interactions among people. In science, the heritage is more through ideas and through phenomena than it is through objects, and so techniques to display ideas and the experiences of nature predominate."[7]

Science and technology centers usually are designed to be "do-it-yourself" museums. They rely heavily on mechanical, electronic, audiovisual, and other exhibit techniques to communicate information. Pushing buttons, turning cranks, lifting levers, and other interactive methods are commonly used to attract and involve the museum visitor. Other times, the participation is emotional or intellectual without the use of physical devices.

Children and adults have found science and technology centers to be extremely enjoyable learning experiences, and educators also have applauded the participatory exhibit approach as an effective supplement to formal classroom instruction. As a result, both the number of science centers and their attendance continue to increase at a rapid pace. In 1981, there were about thirty comprehensive science and technology centers with an annual attendance of approximately 25 million. In addition, at least seventy other museums with an attendance of 15 million were at least partially science centers or were moving in that direction. Although these numbers are relatively small in comparison to the more than 5,500 American museums listed in *The Official Museum Directory,* science and technology centers are among the most popular and effective. A 1974 survey of 1,820 museums conducted by the National Endowment for the Arts showed that 38 percent of all museum visits were to science museums (including science centers, natural history museums, and other science-oriented museums), 24 percent to history museums, and 14 percent to art museums.[8] A 1979 survey by the Institute of Museum Services showed an even higher science museum attendance: 45 percent of all museum visits were to science museums, as opposed to 24 percent to history museums and 12 percent to art museums.[9]

Several decades ago, the words "science museum" were used to describe museums that were involved primarily with natural history and historical collections. Today, because of the growth and popularity of science and technology centers, they normally refer to museums dealing with the natural and applied sciences in a contemporary framework.

In 1964, Bernard S. Finn, curator of electricity at the Smithsonian Institution, questioned the use of the term "museum" to describe the emerging science and technology centers. In the strict sense of the word, he said, it probably did not properly cover these institutions. "... Some, like the Palais (de la Découverte) in Paris and the Institute for Industry and Technology in Amsterdam, purposely avoid using it. Others have preserved the word from earlier days of different aims. Among the curators, opinion seems divided: some feel that the term is sufficiently descriptive and at the same time suggests a certain desirable status; others feel that the term is not descriptive and ought to be avoided because it connotes the dusty, dull, and dreary traditional museum of science and technology."[10]

It was such thinking that led to the use of other words to describe some contemporary museums of science and technology, such as the Buhl Planetarium and Institute of Popular Science, Pacific Science Center, Lawrence Hall of Science, the Exploratorium, Center of Science and Industry, and the Evoluon.

Writing in the UNESCO quarterly review, *Museum,* in 1967, W. T. O'Dea and L. A. West explain:

The concept of a science museum has altered dramatically in the last few years. No longer is it regarded as a collection of historical objects . . . there has been nothing approaching the revolutionary thought that has been given to science museums where the movement is spreading rapidly to integrate them into the educational systems of the countries supporting them.

The great national museums that have been established during the past two centuries still continue to have a preponderance of historical exhibits, perhaps 70% or more, but in the newer museums that are coming along the relative percentages of modern and historical exhibits tend more to be reversed, and it would probably be rare for more than 30% to be of the historical kind. This change is so marked that there is a growing tendency to refer to the new museums . . . as centers of science and technology.[11]

In stating the rationale for this new type of science museum in 1968, Frank Oppenheimer, who later founded the Exploratorium in San Francisco, said that

There is an increasing need to develop public understanding of science and technology. The fruits of science and the products of technology continue to shape the nature of our society and to influence events which have world-wide significance. Yet, the gulf between the daily lives and experience of most people and the complexity of science and technology is widening.

Remarkably few individuals are familiar with the details of the industrial processes involved in their food, their medicine, their entertainment, or their clothing. The phenomena of basic science, which has become the raw material of invention, are not easily accessible by the direct and unaided observation of nature, yet they are natural phenomena which have, for one segment of society, become as intriguing and as beautiful as a butterly or flower.[12]

A similar sentiment was expressed in 1966 when the N. V. Philips electronics organization established the Evoluon in Eindhoven, The Netherlands, as part of its seventy-fifth anniversary observance. The science center's approach was described as follows:

The Evoluon gives a view of a world in constant metamorphosis. In the last couple of years, particularly, changes take place with increasing speed and are far from ended. We are still, stumbling and recovering, seeking better means of controlling the technology to which we gave being.

We seek ways to fit technology into our existence. With the help of technology we have created fresh opportunities, and at the same time, brought fresh problems on our heads. Problems that must be solved. As in the past, science and technology will be called upon.[13]

Loren D. McKinley, former executive director of the Oregon Museum of Science and Industry in Portland, emphasized the educational role of science and technology centers in writing that "museums are traditionally considered neither fish nor fowl. Educators are aware that they exist. . . . The real world considers museums as educational elitists—people who are mainly interested in stockpiling knowledge, rather than finding uses for it. The only real justification for a museum's existence is its awareness and ability to deal with problems of the community. In terms of the practical world of industry, work, the home, and the community, a museum acts as a bridge, helping to translate the art of the educator to those outside the formal educational system. The reverse of this is also true."[14]

When the Association of Science-Technology Centers (ASTC) was founded in 1973, an attempt was made to delineate a science and technology center. The association's bylaws stated that ASTC was created to serve "museums dedicated to communicating to the public a better understanding and appreciation of science and the derived technology."

Member institutions, the bylaws specified, should be "committed to the development of participative exhibits, demonstrations, and programs in the basic sciences and their technological applications. They should enable the general public to gain a first-hand experience with natural phenomena and man-made devices; serve as a resource for formal educational programs; and foster interest in scientific, engineering, and industrial careers, particularly among young people."[15]

Comprehensive science and technology centers were described as those nonprofit museums with substantial exhibits, demonstrations, and programs designed to further public understanding and appreciation of science and technology; that are interdisciplinary in nature, with emphasis on the physical and life sciences; that make use of visitor-participation techniques; that are involved in extensive educational activities; and that meet certain minimum size, budget, and attendance criteria.

In 1978, the ASTC membership criteria was modified accordingly:

Members of the Association of Science-Technology Centers should be major institutions which have substantial emphasis and commitment in two fundamental areas—"science and technology" and "informal education." "Science" is broadly defined to include mathematics, physics, chemistry, biology, natural sciences, anthropology, medical sciences, etc., and the related technologies. The institution's involvement with "informal education" must be more than token and is expected to include the use of interactive exhibits and experiential learning programs in a museum setting.[16]

Science and technology centers always have been accepted as members of the American Association of Museums, but they had difficulties in becoming accredited when the AAM accreditation program began in 1970. Under this program a museum was defined as "an organized and permanent nonprofit institution, essentially education or aesthetic in purpose, with professional staff, which owns and utilizes tangible objects, cares for them, and exhibits them to the public on some regular schedule."[17] Many science and technology centers do not own and care for objects in the usual sense. Planetariums, art centers, and a number of other types of museums also did not fit the definition.

After several years of confrontation and negotiation with the Association of Science-Technology Centers, the AAM accreditation commission modified the definition and accreditation criteria in 1975 to include any science and technology center and institution that "maintains and utilizes exhibits and/or objects for the interpretation of scientific and technological information."[18] Accredited science and technology centers now are treated no differently than the more traditional museums.

A Critical Look at Science and Technology Centers

In bemoaning the state of education and interpretation in art museums in 1939, Francis Henry Taylor told the American Association of Museums annual meeting that

The science museums, on the other hand, have already begun to show the way to remedy this state of affairs and have, in fact, indicated the role of the future. They have been aided in this respect by the depression. For, while we in the art galleries have been squeezing every penny to buy the treasures of European collections in a rapidly falling market, the scientific and industrial institutions have put their money into reinstallation and interpretative exposition. They have not been above employing all of the mechanical aids to learning which are at hand and have pursued the possibilities of new techniques with tireless energy. In brief, they have at last come to a willingness to contemplate their resources in the light of their potential usefulness to society.[19]

Differing views have been presented by others who are scornful of science and technology centers. For example, Duncan F. Cameron, while director of the Brooklyn Museum, wrote: "Our museums are in desperate need of psychotherapy. There is abundant evidence of an identity crisis in some of the major institutions, while others are in an advanced state of schizophrenia."[20] Cameron was especially critical of the Ontario Science Centre in Toronto, where he served as a consultant during the planning stage: "The Ontario Science Centre is certainly not a museum, although it was originally planned as one. Today it contains a veritable chaos of science exhibits mixed with industrial and technological exhibits sponsored by corporations. There is an infinite number of buttons to push and cranks to turn. Interspersed among all of these are hot-dog stands and purveyors of soft ice cream in a claustrophobic maze of cacophonous noncommunication."[21]

One of the strongest condemnations of science and technology centers came from George Basalla, associate professor of history of science and technology at the University of Delaware in 1973. He assailed the "technological progressivism" of contemporary science and technology museums and the "huckster ethos and uncritical spirit of the older international industrial exhibitions [that] are still with us."[22]

One result of this uncritical belief in technological progressivism is that technical museums often tend to be a mixed bag. Rarely are they satisfied to concentrate on a given historical period; instead they mix historical exhibits with some glimpses of "what-we-have-now" or "how-we-do-it-now." Of course, what is implied is that the current processes or the modern machines represent the finest in man's history. At times technical museums seem to have adopted the General Electric slogan as their own: "Progress is our most important product."

Technological progressivism is not only bad history; it fosters an outlook that is all the more dangerous to society because it takes the appearance of a truism when in fact it is a highly debatable proposition. What I have in mind is the assumption that technological progress and social progress are inexorably linked, that a society based on 500-horsepower engines is inherently better, and its citizens are happier, than one based on 20-horsepower engines. This formula equates cultural superiority and personal satisfaction with the conquest of nature by more and more complex and powerful machines. That formula, devised by ethnocentric industrial nations who chose to use the machine as their cultural yardstick, is the working philosophy of too many technical museums. . . .

Visits to some of the better-known technical museums in the United States have led me to conclude that once the machine has been lifted out of its social and physical setting, it is exhibited in one of three different ways: cornucopia, aesthetic object, or item of romantic, sentimental, or humorous interest. Each of these introduces its peculiar distortions of technology, its historical development, and its meaning to society.[23]

In his recent book on the history and functions of museums, Edward P. Alexander, a historian and educator and former director of the interpretation program at Colonial Williamsburg, was critical along the same lines:

The tendency everywhere today is to begin with present machines and technological progress and to show how they operate and the scientific principles on which they are based without much attention to their historical development; to say nothing of the society that produced them. Only a few of the oldest, largest, and best-supported museums collect historical industrial objects. Most science centers put more emphasis on mockups, models, graphs, and multimedia devices. This approach of "presentism" often leads the museum to drop all attempts at study and research; if industry is called upon to design and build the exhibits, curators may be entirely dispensed with, so that impartial and scientific study disappears, and the emphasis is placed on the idea that progress automatically follows technology and on glossy, soft-sell advertising for industry.

Industrialization and the machine have, of course, brought much progress; a large portion of humankind no longer works from sunup to sundown to obtain the bare necessities of life. But industrialization also creates problems—harm to the environment and ecology, neglect of social, cultural, and humanistic values, depletion of resources, and even threats of human extinction. Thus, progress needs to be considered critically—from a holistic social and humanitarian point of view. . . .[24]

Criticism also has come from the Center for Science in the Public Interest, a Washington-based activist group, which has assailed science and technology centers for their industry-sponsored exhibits and for having "succumbed to the influence of corporate dollars by completely abandoning curatorial control.[25]

As citizens seek information about the scientific debates that have become front-page news, science museums have proved to be a highly attractive means of promoting learning in an entertaining milieu. Countless children receive their early exposure to science through the engaging, flashy, and informative displays in museums. What is virtually absent from these shiny displays of technological hardware is any concrete discussion of environmental pollution, consumerism, food safety, small-scale technology, and the role of effective governmental regulation in protecting the public.[26]

The center's *White Paper on Science Museums* makes ten recommendations, including encouraging corporations to make general contributions rather than support specific exhibits, having museums assert more control over exhibit content and credits, and reorganizing governing boards to add representatives from local environmental, consumer, and academic groups.[27]

Criticisms such as these have some merit, but they are grossly exaggerated. They deal more with the past than the present and tend to reflect the views of traditional object-oriented curators, academic historians, and consumer activists on the role of contemporary science and technology museums.

Most science and technology centers did not come into being to serve as repositories for artifacts, to trace the historic development of science and technology, to point out the failings of science and industry, or to function as instruments for self-appointed watchdog groups. They were founded by philanthropists, governments, and community leaders primarily to explain scientific principles, technological applications, and/or industrial operations—areas that were virtually ignored by traditional museums. They have sought to further public understanding and appreciation of science, technology, and industry rather than to pick them apart.

Science and technology centers are far from being perfect creations. They cannot possibly cover everything; they may look favorably upon science and industry; they may not present exhibits and programs that please everyone at all times; they may occasionally overlook or misinterpret a subject; and they undoubtedly have other real or imagined faults. However, there is nothing as effective as a contemporary science center in stimulating interest, communicating information, and entertaining the public in the fields of science, technology, industry, and health. They have achieved this position by focusing on the present and future rather than the past; by emphasizing enjoyable participatory techniques; and by being imaginative, flexible, and persistent in their furtherance of public science education. They will continue to evolve, improve, and develop new approaches as they respond to society's changing needs.

2

The Contemporary Movement

The Beginning

Contemporary science and technology centers have evolved over several centuries from old-line technical museums, international exhibitions, and changing museum concepts and techniques in communicating scientific and technological information to the public.

Most of today's science centers have been founded since 1960, but the concept of furthering public understanding of science and technology through museum exhibits was spawned before 1700, implemented in the 1800s, and refined in the 1900s.

Eugene S. Ferguson, who conducted a study of technical museums in 1965 while professor of mechanical engineering at Iowa State University, credited the international exhibitions of the nineteenth century with the greatest impact, stating: "The idea of a technical museum has firm roots in 17th and 18th-century cabinets of mechanical models; the idea was enlarged in the 19th century fairs and exhibitions of mechanics institutes; but with few exceptions today's technical museums owe their existence to the international exhibitions of the 19th century."[1]

Ferguson may have been right about technical museums in 1965, but a majority of the new science and technology centers are largely a response to local, regional, or national science education needs. Francis Bacon, the English philosopher and statesman who was one of the earliest and most influential supporters of empirical and scientific methods, proposed the establishment of a museum of inventions and a gallery of portraits of inventors about 1600 to point out the practical importance of the emerging mechanical arts and sciences.[2]

Later in the seventeenth century, René Descartes, the French philosopher, mathematician, and scientist, proposed a museum containing scientific instruments and the tools of mechanical trades. He urged that a skilled artisan or mechanic be attached to each trade group to answer questions regarding processes and the use of tools. The plan was not put into effect, but it influenced the establishment of the Conservatoire des Arts et Métiers (National Conservatory of Arts and Trades) in Paris in 1794.[3]

The French Académie des Sciences (French Academy of Sciences), shortly after being founded in 1666, began collecting models of inventions submitted by those who wished to receive the academy's endorsement of their developments. A seven-volume descriptive catalog of the models later was published by the society.[4]

In 1675, Gottfried Wilhelm Leibniz, the German scholar, mathematician, and philosopher, advocated the establishment of an exhibition or museum of machines and other inventions to enlighten and entertain the public. At this time the only technical "museums" were private collections of mechanical models, instruments, and curiosities of nature.

Leibniz proposed that the exhibits include "Magic Lanterns . . . flights, artifical meteors, all sorts of optical wonders; a representation of the heavens and stars and of comets; a globe like that of Gottorp at Jena; fire-works, water fountains, strangely shaped boats; Mandragoras and other rare plants. . . ."[5]

He also suggested a demonstration "to show how a child can raise a heavy weight with a thread," "new experiments on water, air, vacuum," tests for machines "which would throw things exactly at a given point," a telescope to "show the moon at night along with other heavenly bodies," "exhibits of the muscles, nerves, bones," and other exhibits that are commonplace in contemporary science and technology centers.[6]

Leibniz stated that such a undertaking "would open people's eyes, stimulate inventions, present beautiful sights, instruct people with an endless number of useful or ingenious novelties. . . . It would be a general clearinghouse for all inventions, and would become a museum of everything that could be imagined."[7]

The words of Leibniz went unheeded, but a few years later, in 1683, the world's first museum—the Ashmolean Museum—was founded at Oxford University in England. The Museum grew out of the natural history collection of Elias Ashmole and was the forerunner of Oxford's Museum of the History of Science, which took its present form in 1949. Although narrow in scope, the science history museum has one of the most impressive collections of antique scientific instruments in the world.

Europe has at least three other noteworthy museums dealing with the history of science—all of more recent vintage. Many of their contents, which are primarily instruments of science rather than machines, go back to the Renaissance and the centuries that followed. The ancient instruments, such as Galileo Galilei's telescopes, exhibited at the Instituto e Museo di Storia della Scienza (Museum of the History of Science) in Florence, Italy, came from the collections of the Medici family. Many historic scientific instruments and apparatus also can be seen at the Het Rijksmuseum voor de Geschiedenis der Natuurwetenschappen (National Museum for the History of Science) in Leyden, The Netherlands, and at the Museum voor de Geschiedenis van de Weterschappen (Museum for the History of Science) in Gent, Belgium.

In 1695, Christopher Polhem, the father of Swedish technology, suggested that a permanent display of machines be included in a proposed Bureau of Mines laboratory for developing useful machines. Although the idea was not implemented, the Swedish Royal Model Chamber did come into being in 1748 when Polhem provided models of a number of his machines for exhibit in the Royal Palace.[8]

With the advent of the Industrial Revolution came a number of societies to encourage inventive and industrial development. The first of these organizations was "the Society for the Encouragement of Arts, Manufactures, and Commerce" (now the Royal Society of Arts) in England in 1754. In 1761, the society provided a guide to explain to the public the collection of models gathered from its various competitions. Although many of the models later were destroyed or dispersed, many eventually were turned over to the Science Museum in London.[9]

With the backing of Benjamin Franklin, a similar society was established in Philadelphia in 1766. Two years later, "the American Society for Promoting and Propagating Useful Knowledge" merged with the American Philosophical Society. One of the activities of the organization was a collection of models.[10]

Early Technical Museums

At the end of the eighteenth century, a slightly different approach to furthering inventive and industrial development was launched in France as an outgrowth of the French Revolution. In 1783 Jacques de Vaucanson, a French inventor and mechanic, bequeathed his extensive collection of machines, instruments, and models to King Louis XVI of France. The King decided to place the collection on exhibition as a means of explaining their operations to the public, instructing artisans, and inspiring new innovations by inventors.[11] In 1794 the Conservatoire National des Arts et Métiers was created as a teaching institution for the applied arts and sciences, and the Vaucanson collection was used at the conservatory to demonstrate design principles and practices. The collection also became the basis for the first technical museum, the Musée National des Techniques.

In the nineteenth century great quantities of scientific and technical equipment were added, with the most significant additions being in 1814 and 1866. They included artifacts from such leading French scientists as Jacques Alexandre César Charles, Antoine Laurent Lavoisier, and Charles A. de Coulomb. In 1819, public courses on applied science in the arts and industries were instituted.

The Musée National des Techniques, however, failed to keep pace with changing times, largely because of the lack of adequate funding and imaginative leadership. Few additions were made to the collections and almost no changes were made in operations after 1900. As a result, a museum that was at the forefront of scientific and technological developments at its founding became a neglected storehouse of eighteenth- and nineteenth-century artifacts. Among its collection of 80,000 historic objects and models are the first steam-powered carriage by Nicholas Joseph Cugnot, the first calculator by Blaise Pascal, and the first steam-powered airplane by Clément Ader.

During the 1820s, mechanics' institutes were formed in the United States and Great Britain to provide technical training and a showcase for industry. The Franklin Institute of the State of Pennsylvania for the Promotion of the Mechanic Arts, for example, was established in 1824 when an ambitious young man, Samuel V. Merrick, found to his dismay that no means existed to obtain mechanical training unless he became indentured to someone already in the business.[12] The institute offered classes in subjects such as mineralogy, chemistry, natural philosophy, mechanics, architecture, drawing, machine design, and mathematics, and public lectures were given on the progress of science and technology by leaders in American science, invention, and industry.

Concerned with the development of the sciences and their practical applications, the Franklin Institute also decided to organize an industrial exhibition and offer prizes for meritorious examples of products. An American Manufactures Exhibition was held in Carpenters' Hall in October 1824 and was so successful that it was followed by some twenty-five similar exhibitions over the next thirty-four years.[13]

As a result of these experiences, a permanent cabinet of models and minerals was formed in the society's library. Approximately 100 years later, a science museum and planetarium were added to the Franklin Institute as part of its expanded efforts to further public understanding of science and technology.

Although trade and national exhibitions became common, it was not until 1851 that the first major international exhibition was held. The Exhibition of the Industry of All Nations, better known as the "Crystal Palace Exhibition" or the "Great Exhibition," was presented by the Royal Society of Arts in London to stimulate and promote British industry but included representation from throughout the world.[14]

The focal point of the Great Exhibition was a huge iron and glass building—the Crystal Palace—that covered eighteen acres and was constructed of 200,000 pieces of glass. Unlike most of its successors, the Great Exhibition was financially successful and produced a surplus of £180,000. With the funds collected and materials salvaged from the exhibition, the South Kensington Museum of Industrial Arts (later renamed the Victoria and Albert Museum) was opened in 1857.[15]

In 1909, science collections were separated from the decorative art collections to form the Science Museum, the national museum of science and technology in Great Britain. Although the Science Museum inherited miscellaneous mechanical materials from the Exhibitions of 1851 and 1876, some of its most significant collections were obtained from the Patent Office, which, under Commissioner Bennet Woodcroft, notably rescued James Watt's rotative steam engine of 1788 and George Stephenson's *Rocket* locomotive of 1829.[16]

In the years that followed, the Science Museum became one of the great technical museums, with perhaps the richest collection of scientific and technological artifacts. Although some of its machines could be demonstrated, it remained object-oriented, and it was not until after World War II that a "Children's Gallery" with participatory exhibits was added to its offerings.

Many international expositions were held in the half-century following the Great Exhibition of 1851, some of which inspired the founding of other technical museums in other countries. For example, the 1862 International Exposition in London influenced the formation of the Bohemian Industrial Museum, which later became the National Technical Museum in Prague; the 1873 International Exposition and the 1908 Austrian Exposition produced the Technological Museum of Industry, Crafts, and Trades in Vienna; the 1876 Centennial International Exhibition in Philadelphia helped establish the Smithsonian Institution's National Museum and its Department of Arts and Industries in Washington; and the German Electrical Exposition of 1882 and the succeeding International Electrical Exposition at Frankfurt-on-the-Main led to the founding of the Deutsches Museum in Munich.

The purpose of most of these scientific and industrial institutions was "to increase the means of industrial education and extend the influence of science and art upon productive industry" and "to enlighten the people by exposing them to the fruits of technical progress," according to F. Greenaway and Bernard S. Finn, respectively.[17] At the same time, pointed out George Bassala, "these 'tournaments of industry' pitted the industrial nations against one another in friendly rivalries, celebrated the machine and its various products, and—above all—gave prominence to the notion that quantifiable technical progress was to be equated with social progress and human betterment."[18]

Among the museums that benefited from international expositions was the U.S. National Museum, which was operated by the Smithsonian Institution in Washington. The museum's collections were enhanced by twenty-one railroad freight cars of natural history, scientific, technological, and industrial residue from thirty countries from the Centennial International Exhibition of 1876 in Philadelphia. These objects were salvaged by G. Browne Goode, a young curator who later became director of the National Museum. They formed the basis of the museum's new Department of Arts and Industries, which was incorporated into the Department of Anthropology in 1897 following Goode's death. Goode was ahead of his time in many ways, as evidenced by his paper on museums of the future: "The museum of the past must be set aside, reconstructed, transformed from a cemetery of bric-a-brac into a nursery of living thoughts. The museum of the future must stand side by side with the library and the laboratory, as part of the teaching equipment of the college and university, and in the great cities cooperate with the public library as one of the principal agencies for the enlightenment of the people."[19]

A New Approach

The first of the new-style science and technology museums was the Deutsches Museum von Meisterwerken der Naturwissenschaft und Technik (The German Museum for Masterworks of Natural Science and Engineering), which became known simply as the Deutsches Museum, in Munich. For twenty years Oskar von Miller, a leading German electrical engineer who was involved in organizing two electrical expositions, had toyed with the idea of founding a new type of science and technology museum in Germany. In addition to historic artifacts, he wanted to include working sectioned models, demonstrations, and visitor-participation devices to illustrate scientific, engineering, and industrial history and principles.

In 1903 he formed a foundation for the purpose of implementing the idea. His proposal was received with enthusiasm by government, scientific, and industrial leaders. He succeeded in convincing Prince Ludwig Von Bayern, later King Ludwig III, to become patron of the foundation and enlisted leading scientific and industrial figures, such as Carl von Linde, Walter von Dyck, Wilhelm von Siemens, Conrad Roentgen, and Anton von Rieppel, to serve on the board.

In 1906 the provisional collections were opened to the public in the former National Museum, and supplemental exhibits were added in the old Isar Barracks in 1909. The Deutsches Museum's first building on its present site was completed in 1925. It was supplemented with a library building in 1932 and the Congress Hall in 1935. Most of the museum complex was destroyed by Allied bombs during World War II, and it was not until 1970 that virtually all of the museum was restored.[20]

"The German museum in Munich," von Miller explained, "shows the development of various branches of natural science and technology by means of original apparatus and machines, as well as by means of models and arrangements for demonstration, in a manner easily understood by all classes of people. Its purpose is to instruct students, workers, etc., as to the effects of the multifarious applications of science and technology to the problem of human existence, to stimulate human progress, and to keep alive in the whole people a respect for great investigators and inventors and their achievement in natural science and technology."[21]

The Deutsches Museum contained many historic objects of significance, such as a 1588 astrolabe by Erasmus Habermel, the 1663 vacuum globe and air pump by Otto von Guericke, a 1779 reflecting telescope with viewfinder by G. F. Brander, the first dynamo (1866) made by Werner von Siemens, the first motor car (1886) by Carl Benz, the 1895/1896 X-ray discharge tubes used by Wilhelm Roentgen, and the first diesel engine (1897) by Rudolf Diesel.[22] But it was von Miller's use of new exhibit techniques—such as operating models, full-size machine replicas, a walk-through coal mine, science demonstrations, a cutaway submarine, and exhibits that could be activated by visitors—that captured the imagination of the public and the attention of other museums.

"Experimental exhibits intended to educate the public in a museum," von Miller pointed out, "must be designed with the greatest simplicity. They are operated by the average inexperienced museum visitor, or, at least, by the guards. They must be built sturdily, with the fragile parts protected from handling, and must produce the desired results quickly and often continuously. The results must be obvious, so that they can be easily and clearly observed."[23] His advice still holds true today.

Two other influential technical museums were founded in Europe shortly after the Deutsches Museum began operations. The Technical Museum of Bohemia, which later was reorganized and renamed the Národní Technické Museum (National Technical

Museum) was established in Prague in 1908, and the Technisches Museum für Industrie und Gewerbe (Technical Museum of Industry, Crafts, and Trades) was created in Vienna in 1909 and opened in 1918.

Although collections were used in Bohemian technical training classes as early as 1799, it was not until 1862 that the first industrial museum was formed. Called the Bohemian Industrial Museum, it exhibited many objects purchased from the London and Paris expositions. It was followed by the Museum of Applied Arts in 1885, the Technological Museum in 1898, and finally the National Technical Museum of Bohemia. The bulk of objects for the technical museum came from the earlier museums and an industrial and trade show in Prague in 1908.[24] It was largely historical in nature and did not follow the example of the Deutsches Museum.

Wilhelm Exner, the founder of the Technical Museum of Industry, Crafts, and Trades in Vienna, however, attempted to improve upon the floor plan and exhibit presentation of the Deutsches Museum, despite financial and other difficulties. When completed, the museum's main floor formed one large exhibition space with only a few partitions, which made it possible to display the collections, working models, and participatory exhibits in a much more inviting setting. The second floor was devoted mainly to an explanation and demonstration of various process industries, usually from a historical perspective.[25] With the passage of time and lack of adequate funding, the technological museum in Vienna became primarily a repository of historic artifacts, including the sewing machine of Joseph Madersperger, the first metal camera by Friedrich Voigtlander, the first calculated portrait lens by Joseph Petzval, the ship propeller by Joseph Ressel, the low pressure turbine by Viktor Kaplan, and the electric power wheel by Johann Kravogl.[26]

Despite the development of technical museums in Paris, London, Prague, Vienna, and other cities, it was the Deutsches Museum in Munich that became the prototype for the new science and technology museums in other parts of the world. Although basically historical and object-oriented like the others, its use of new participatory and demonstration techniques helped to change the nature and direction of technical museums.

American Adaptations

The increasing number of technical museums in Europe resulted in calls for similar institutions in the United States. Among those urging the founding of American technical or industrial museums was Charles H. Richards, a former director of the American Association of Museums and author of *The Industrial Museum* (1925).

"We are today one of the foremost industrial countries of the world," he stated. "Can we afford to omit from our educational program the story of what has made us? We have a high type of industrial organization and as a people we are the first to utilize the fruits of new inventions. Shall we leave other nations to grow wise through the study of our achievements and ourselves neglect their meaning and their inspiration? To tell the story adequately we need the industrial museum."[27]

The first large-scale museum of science and technology to open in the Western Hemisphere was the Henry Ford Museum, which was launched with great fanfare in Dearborn, Michigan, in 1929. Founded by Henry Ford, the museum and the adjacent outdoor historical museum, Greenfield Village, were dedicated to preserving America's heritage.[28]

The museum reflected the collecting philosophy of its founder. It consisted of three parts under one huge roof. At the front—housed in reproductions of Philadelphia's Independence Hall, Carpenter's Hall, and the Old City Hall—were the American Decorative Arts galleries, with outstanding collections of furniture, ceramics, glass, pewter, silver, and textiles from the Pilgrim period to the late nineteenth century. At the rear was the great Mechanical Arts Hall, with extensive collections relating to agriculture, home arts and crafts, industrial machinery, steam and electric power, lighting, communications, and transportation. Connecting these two major sections was the Street of Early American Shops, illustrating America's eighteenth- and nineteenth-century craft shops.[29]

Over the years, the museum's collections have been expanded but its historial orientation has remained virtually the same. As explained in its guidebook, "the Henry Ford Museum is a general museum of American history housed in the most American of buildings. Here, with collections matchless in both depth and scope, is the story of America's growth, its inventive genius, the development of its tastes and modes. Any one section of this Museum, be it agriculture, power, transportation, communications, or furniture, is a fascinating history of the American people—their yearnings, their explorations, their accomplishments."[30]

The first of the contemporary science and technology museums in the United States was the New York Museum of Science and Industry. Opened in 1930, it traced its origin to the formation of "The Association for the Establishment and Maintenance for the People in the City of New York of Museums of the Peaceful Arts" in 1914.[31] Among those supporting the concept were Thomas Alva Edison and Henry R. Towne, president of the Yale and Towne Manufacturing Company. When Towne died in 1924, he established a provisional endowment fund to help support a "peaceful arts" museum. The City of New York also directed that fifteen acres of land at the site of the old Jerome Park Reservoir might be provided for the museum.

In 1926, a task force headed by Charles T. Gwynne, a member of the museum association's executive committee and executive vice president of the Chamber of Commerce of the State of New York, went abroad to study and make a motion picture of the new great industrial museums in Europe. The completed study report and film were titled *Museums of the New Age.*

"In America," the report stated, "popular conception of a museum is unfortunately not of an educational institution ranking with a college or a university. It is rather of a place of curiosities, of nine-day wonders, of freaks of nature, of love's labor lost in engraving the Lord's Prayer on the head of a pin and that sort of thing."[32]

The study team was impressed with what it saw in the European museums' efforts to popularize science and technology. "The first different impression that one obtains in visiting the industrial museums . . . is that exhibits move," the report pointed out. "Indeed, on a visiting day in almost any of the institutions concerned, there is a distinct sense of life."[33]

The report further stated that "the second most remarkable thing about these exhibits . . . [is] that visitors there *handle the exhibits.* Handling points to the basic departure of new from old. There has been a radical change in educational approach. Those in charge today realize that 'knowledge is doing'—the profoundest pedagogical principle ever enunciated. This principle, in successful practice, is the real outstanding possession of these amazing 20th century institutions."[34]

The New Yorkers also were impressed with the popularity of these new museums and their exhibits. "It was remarkable, too, that in these industrial museums, while obviously white-collar workers were much in evidence among the visitors, there was a strong percentage of men and women who undoubtedly labored with their hands," according to the report.[35]

The study team recommended the establishment of an industrial museum that was "primarily educational rather than primarily historical." In 1930, the New York Museum of Science and Industry opened as a section in the lobby of the New York Daily News Building. The museum struggled along until 1935 when plans were announced to expand and relocate the museum in Rockefeller Center. A fifteen-year lease was signed for the 50,000 square-foot Forum Exhibition Hall in the center's RCA Building, and in 1936 "the rudely intricate devices of the dawn of the industrial revolution, the delicately powerful developments of modern experimental science, and the fundamental forces which link them" awed the guests at a glittering preview.[36]

The New York Museum was called the "Museum of Motion" because of its more than 400 exhibits that "are in operation or can be set in operation by the visitor by the press of a button, the turn of a crank or the pull of a lever." The exhibits were arranged in eleven divisions—food industries, clothing-textiles, shelter, highway transportation, railroad transportation, aviation, communication, machine tools, power, and electrical science and technology.[37]

The museum, however, continued to have financial and operating problems—despite the assistance of the Carnegie Corporation, Rockefeller Foundation, and others. When its lease expired at Rockefeller Center in 1950, the museum moved to a smaller space on the second floor of the Hotel Claridge under the name of the New York Hall of Science (no connection with the Hall of Science of the City of New York that opened in 1966), but it went out of business a few years later from the lack of interest and support.

As it turned out, the second institution to open—the Museum of Science and Industry in Chicago—is now the oldest, largest, and best attended of the nation's contemporary science and technology museums. The museum was founded by Chicago businessman and philanthropist Julius Rosenwald as a result of his eight-year-old son's fascination with the exhibits, and particularly the machines that could be activated, at the Deutsches Museum during a 1911 family vacation in Munich. Rosenwald went to the museum to find out what was so intriguing about the place and came away convinced that a similar museum should be established in the United States.

A decade later, Rosenwald, who was chairman of Sears, Roebuck & Company, had an opportunity to pursue the idea. It was at that time that Chicago was debating what to do about the deteriorating classic Greek structure that served as the Palace of Fine Arts at the 1893 World's Columbian Exposition and later was used to house the Field Museum of Natural History until its own building was completed. Rosenwald felt the building could be rehabilitated to make an excellent home for this new type of museum dealing with the mechanical arts.

In a letter to Samuel Insull, then president of the influential Commercial Club of Chicago, Rosenwald wrote: "I have long felt that Chicago should have as one of its important institutions for public usefulness a great Industrial Museum or Exhibition, in which might be housed, for permanent display, machinery and working models illustrative of as many as possible of the mechanical processes of production and manufacture."[38]

He explained that such a museum should be created "for the entertainment and instruction of the people; a place where workers in technical trades, students, engineers, and scientists might have an opportunity to enlarge their vision; to gain better understanding of their own problems by contact with actual machinery, and by quiet examination in leisure hours of working models of apparatus; or, perhaps, to make new contributions to the world's welfare through helpful inventions. The stimulating influence of such an exhibit upon the growing youth of the city needs only to be mentioned."[39]

Rosenwald offered to provide $1 million for the exhibits if the Commercial Club would spearhead the project and funds could be found to rebuild and adapt the structure. The Commercial Club endorsed the concept and the South Park District floated a $5-million bond issue to rehabilitate the building. The museum was incorporated in 1926 and opened in 1933. Meanwhile, reconstruction of the

600,000-square-foot building and expansion of the exhibits continued through 1940, with the assistance of an additional $6 million from Rosenwald.

In writing about the 1933 opening of the Chicago museum, the first director, Waldemar Kaempffert, said:

What the museum presents today is a cross-section of what it will ultimately become, something which reveals the principle of letting operating mechanism teach the facts of science, engineering, and industry in the only way in which they can be taught vividly and interestingly. All the art of the stage director is invoked to tell a technical story. Yet, this is no technical Coney Island. The dramatic is introduced not for its own sake, but as a means to an educational end.

It is not the purpose of the museum to furnish entertainment, but if flashing machines hold the attention, if charts and dioramas fascinate because lights flicker within them to bring out a point and the visitor has thereby learned something, the means will be justified. There is nothing so dramatically instructive as action. Crowds will watch a machine produce cigarettes by the thousands an hour, but no will give it more than a passing glance if it stands idle. The museum follows the tendency to vivify teaching.[40]

Although the building was incomplete, the Museum of Science and Industry was opened in 1933 to coincide with the Century of Progress Exposition in Chicago. The museum's staff helped to prepare the basic science exhibits at the exposition with the understanding that the exhibits would come to the museum after the close of the exposition. The museum also inherited numerous industrial exhibits from the fair. The museum's exhibits covered many fields, including agriculture, rail transportation, electrical power, communications, chemicals, machinery, and minerals. Its most popular exhibit was a full-scale coal mine, which continues today after numerous updatings. Nearly 300,000 people came to the museum in its first year.

In the depression years that followed, the museum experienced serious financial problems as construction and operating costs increased, the attendance plateaued, few new exhibits were added, and industrial support was insufficient. Rosenwald had died in 1932 and could not bail out the museum.

In 1940 the museum trustees convinced Lenox Lohr, president of the National Broadcasting Company and former general manager of the Century of Progress Exposition, to take the presidency. Lohr believed that the traditional organization of the museum into departments with curators for each subject areas made for inflexibility and high costs. Thus one of his first acts was to dismiss most of the eighteen curators and a number of other staff members as a means of cutting costs and making changes in the exhibits. He initiated an industrial participation program to reverse the 90 percent historical and 10 percent contemporary composition of the exhibits he found upon his arrival. With financial support from industry, he also wanted to change at least 10 percent of the exhibits each year to keep abreast of current developments.

Lohr was successful in obtaining major new exhibits on telecommunications from the Bell Telephone System, abrasives from the Carborundum Company, automotive engineering from General Motors Corporation, farming from International Harvester Company, aluminum from the Aluminum Company of America, plastics from the Bakelite Corporation, and railroads from the Santa Fe Railway. At the same time, he succeeded in eliminating the deficit, dramatically increasing the attendance, and converting the institution from a historical to a contemporary science and technology museum. In the process, it also became a prototype for other science and technology museums.

While the Chicago museum was undergoing birth pains, the Franklin Institute Science Museum and Planetarium was founded in Philadelphia in 1934 and the Buhl Planetarium and Institute of Popular Science was established in Pittsburgh in 1939. The Philadelphia museum was an outgrowth of the Franklin Institute created for the promotion of the mechanic arts in 1824, while the Pittsburgh museum was a gift of the Buhl Foundation in memory of a local industrialist, Henry Buhl, Jr., and his wife, Louise.

Of the two Pennsylvania contemporary museums, the Franklin Institute Science Museum developed more fully and rapidly. In its first century, the Franklin Institute sought to achieve its mechanical science popularization objectives through public lectures, classes, experimental workshops, research programs, publications, and even exhibitions. As times changed, however, the institute was forced to abandon its schools, exhibits, and a number of other programs.

In the search for a new means of diffusing a knowledge of science and its impact upon the lives of people, the board of managers considered in 1922 the possibility of establishing a museum of science and technology that "would fulfill a unique role as a supplement to the education of youth and as an agency for adult education."[41]

The idea was reactivated in 1927 when the Poor Richard Club of Philadelphia proposed that a national memorial dedicated to Benjamin Franklin be incorporated as part of a new museum to be built by the Franklin Institute. Despite the financial collapse of 1929, a civic committee headed by magazine publisher Cyrus H. K. Curtis raised $5.1 million for the project in just twelve days. William L. McLean, a newspaper publisher, made a large contribution for a heroic-size statute of Benjamin Franklin, and Samuel S. Fels, an industrialist and philanthropist, made a sizable donation for a planetarium.

Like other science and technology museums of the period, the Franklin Institute Science Museum was largely historical in orientation. Among the objects on display were Joseph Priestley's air pump, materials used by Michael Faraday in his electromagnetic discoveries, Thomas Alva Edison's three-wire generator first used for street lighting, the apparatus of Elihu Thomson, John A. Fleming's first radio tube, the original Lee de Forest audions, and early television tubes developed by Vladimir Zworykin.[42] The institute also had exhibits presented by industrial firms, participatory exhibits, and a planetarium. "The museum introduced to America a new phase of visual education," an institute historian wrote. "Intended to center around the scientific interests of Benjamin Franklin, it furnished a comprehensive survey of science and its numerous applications in everyday life. It is a virile dynamic assembly of exhibits, the antithesis of the old museums which are aptly described as 'community attics.'"[43]

It was the pioneering industrial museums in New York, Chicago, and Philadelphia that provided the foundation for the more recent evolution of contemporary science and technology centers.

Contemporary Science Education Centers

The vast majority of science and technology museums that have come into being since World War II have de-emphasized historic objects and accentuated participatory exhibits for public science education purposes. Some actually do not have any artifacts. This change can be attributed to a number of factors. A tremendous increase in the number of such institutions was accompanied by a decline in the availability of artifacts. Another factor was the shift from large national museums founded by governments and wealthy benefactors to smaller local and regional centers started by community groups. Even more important was the change in the purpose of the new science and technology museums, as well as a difference in public expectations.

Rather than focusing on the past, most new institutions were more concerned with the present and the future. They sought to communicate current scientific and technological information to the public and do so in an enjoyable and educational manner. Instead of collections of objects, they relied more heavily on constructed exhibits designed specifically for effective involvement and learning.

Science and technology museums became more than "museums." As they assumed greater responsibility for public awareness and understanding of science and technology, they also were transformed into informal and supplemental educational institutions. They became centers of popular science education and instruments for increased public interest and involvement in science policy issues.

Part of this transformation can be attributed to changing public interests, expectations, and needs. The world has become more complex as a result of scientific and technological advances. In general, people are better educated, earn more, and have more leisure time than past generations. There are more demands on their time by television, sports, community affairs, and other activities. When they go to museums, they like to be entertained. They also want to know more about the scientific and technological world around them.

The first of the contemporary science and technology institutions was the Palais de la Découverte (Palace of Discovery), which was launched in Paris as an offshoot of an international exhibition in

1937. The scientific exhibits and demonstrations originally were organized by Jean Perrin for the fair, but the concept of popular science education became so popular that it formed the basis for a permanent institution.

The Palais had virtually no artifacts. It sought to explain scientific principles and technological applications to the general public largely through experimental demonstrations given by University of Paris students. A Smithsonian historian wrote that "The Palais is a more or less permanent exhibition of science rather than a museum of the history of science. The concept of setting up an animated textbook of science—or perhaps one should call it a laboratory extended in space rather than time—has been carried out in varying degrees elsewhere, but never on so grand a scale as here, where one may see animated demonstrations not only of the experiments of Galileo and Newton but also of those of de Broglie (undulatory mechanics) and Langevin (ultrasound)."[44]

Perrin's objectives were "to bring to everybody's attention the progress of science and technology; to develop the scientific spirit and hence the qualities of precision, of honest criticism, of free judgment; to illustrate teaching at all levels; to give to teachers the means of bringing up to date their knowledge; to orient the young towards a career corresponding to their capabilities and interests; to participate in recycling; to enable everybody to adapt themselves to the modern world under the best conditions."[45]

The popularity of the Palais de la Découverte grew as it expanded its offerings. The following describes its mission in 1975: "Theatre, literature, art, sculpture, music, and even sport have always been privileged in their propagation. Science, which nowadays advances in leaps and bounds, needed to penetrate, in its turn, into everybody's daily life, to leave the confines of the laboratory and the school, as it were to make itself known."[46]

Making use of fifty-three exhibit halls, the Palais presents exhibits and demonstrations in seven broad fields—astronomy, biology, chemistry, earth sciences, mathematics, medicine, and physics. It also offers public lectures, science films, temporary exhibitions, field trips, study camps, outreach programs, and the use of its laboratories for research by youngsters between fourteen and eighteen. The Palais is scheduled to be incorporated into the new National

Museum of Science and Industry being developed in Paris by the French government.

In the 1940s, the Museum of Science and Industry in Chicago and the Deutsches Museum in Munich began to change. The Chicago museum de-emphasized its collections, fired its curators, and added a substantial number of participatory constructed exhibits dealing with current science, technology, industry, and health—many with industrial sponsorship. When the Deutsches Museum, which was severely bombed during World War II, began the arduous task of rebuilding that continued over some twenty-five years, it incorporated new do-it-yourself exhibit techniques and put more emphasis on contemporary science and technology.

As part of this movement, the contemporary Nederlands Instituut voor Nuverheid en Techniek (Dutch Institute for Industry and Technology) was founded in Amsterdam in 1954 by industrial leaders with financial assistance from the national government. It sought to familiarize youngsters and the general public with Dutch industry.

The role of science and technology in World War II and later the exploration of space created greater public interest in scientific and technological activities. More communities wanted to have their own science centers. Some started new institutions, while others converted—either partially or entirely—existing museums into facilities featuring science and technology. Various new specialized "museums" came into being in such fields as health, energy, and space. In developing nations, science and technology centers were looked upon as instruments of social uplifting and technological growth.

A number of natural history museums were broadened to include the natural sciences and technology. They included the Boston Museum of Natural History, which traced its origin to 1830; the Cranbrook Institute of Science, founded in 1932 in Bloomfield Hills, Michigan, added a physics wing in 1964; a struggling Portland Museum of Natural History became a broad-based Oregon Museum of Science and Industry in 1954; and the National Science Museum, begun in Tokyo as the "Education Museum" in 1877 and converted to a natural history museum in 1931, added a building in 1978 to house extensive technological exhibits.

Among the other conversions were the California Museum of Science and Industry, which started as a state agricultural exhibit hall in Los Angeles in 1912 and evolved into a science and technology center in 1951; the Dallas Health Museum, founded in 1946, changed its name to Dallas Health and Science Museum in 1958 because of its expanded science program; and the Franklin County Historical Society in Columbus, Ohio, converted a small historical museum into the Center of Science and Industry in 1964.

Long-established children's museums in Brooklyn, Boston, and Indianapolis took steps to strengthen their scientific and technological exhibits and programs. Meanwhile, new types of specialized science centers began to appear—health museums in Rochester, Cleveland, and Dallas; energy museums in Oak Ridge, Albuquerque, and Mexico City; space museums in Huntsville, Alabama, and Jackson, Michigan; and a science center concerned with perception in San Francisco.

The science exhibition halls at two American world's fairs became permanent science and technology centers in the 1960s. The Pacific Science Center was opened in the U.S. Science Pavilion after the 1962 Seattle World's Fair, and the Hall of Science at the 1964–1965 New York World's Fair became the New York Hall of Science.

More than twenty other science and technology centers were founded in the 1960s and 1970s as the pace accelerated: the Nagoya Municipal Science Museum in Japan in 1962; the Evoluon in Eindhoven, The Netherlands, in 1966; Fernbank Science Center in Atlanta in 1967; Lawrence Hall of Science in Berkeley, California, in 1968; Exploratorium in San Francisco and Ontario Science Centre in Toronto in 1969; and Singapore Science Centre in Singapore in 1977.

The Museum of Science in Boston, the Oregon Museum of Science and Industry in Portland, and the Exploratorium in San Francisco, in particular, broadened the science center concept and placed greater emphasis on the educational role.

In Boston, the Museum of Science emerged in the post-World War II years from the Boston Society of Natural History, which was founded in 1830, and its Boston Museum of Natural History, located in an inadequate structure and experiencing attendance and financial difficulties.

The board of trustees decided to hire a young geologist, Bradford Washburn, to run the old natural history museum and to help plan a new museum. Washburn proposed an entirely new type of museum that would encompass all the sciences—the natural, physical, and applied sciences, as well as man and medicine.

Karl T. Compton, then president of the Massachusetts Institute of Technology, reacted favorably but cautioned them: "You have a marvelous idea, but you've got to be brutally selective. If you try to be a museum of *all* science, your building will stretch for miles. And when your plant is a quarter completed, the first half of the exhibits will be obsolete."[47]

Compton's comments led Washburn to the idea of offering a smorgasboard of exhibits with the prime objective being "to open eyes, to broaden horizons, to stimulate curiosity," rather than to be all inclusive and to tell the public everything about science.[48] The result was a combination museum of natural history, the physical and life sciences, and technology, with teaching as its prime objective.

In 1949 the renamed Museum of Science moved to a temporary building on a six-acre site overlooking the Charles River. The land was leased from the Metropolitan District Commission for 99 years at one dollar a year. In the years that followed, two permanent exhibit wings, a planetarium, a theater of electrical science, a six-story central core, and a five-floor garage were built by the Boston science center.

A similar transformation took place on the West Coast about the same time. After ill-fated efforts in 1906, 1930, and 1939 to organize a full-scale Portland Museum of Natural History, a state-wide "Committee of 100" was appointed by Governor Earl Snell in 1943 to develop plans for a museum that would be acceptable to the residents of Oregon. The study committee recommended that a museum be founded to house natural history and historical materials and to serve as a center for educational, industrial, historical, and scientific demonstrations. In 1944 an Oregon Museum Foundation was established to implement the concept.[49]

The museum became the Oregon Museum of Science and Industry in 1954 to reflect its changing emphasis, and in 1957 a drive was launched to build the museum's first permanent home and to broaden its offerings. A site was leased from the City Park Bureau and an old-fashioned "barn-raising" was held, with building suppliers and contractors providing the materials and labor unions and volunteers supplying the manpower. In one day 102,000 bricks were laid by 400 workers in this community effort, and the science center became a reality.[50]

In the years that followed, the Oregon Museum of Science and Industry pioneered a number of educational programs for science centers. OMSI, as the museum became known, established a wilderness summer camp, a "suitcase" school exhibit program, a student laboratory, inner city branch facilities, and an energy education program. It helped to take science and technology centers beyond their walls and into the schools, the neighborhoods, and the wilderness.

The Exploratorium in San Francisco was the brainchild of Frank Oppenheimer, a University of Colorado physics professor who became convinced that museums should take a more active role in furthering public understanding of science and technology. In 1966 Oppenheimer called for a new type of science and technology museum in addressing the role of museums in education. "It is a scandal," he said, "that there are so few museums of science and technology. We live in an age in which science dominates a very large fraction of our efforts and our surroundings. Yet, despite the predominance of technology and the efforts devoted to science, it is not unusual to encounter a strong anti-science attitude in the general population."[51]

Oppenheimer criticized the "passive pedagogy" of most museums and argued for institutions with core participatory materials that would be integrated with the local school system and be effective in explaining scientific phenomena to the public.[52]

In 1968 Oppenheimer put his ideas into practice when he opened the Exploratorium in the Palace of Fine Arts, a building from the 1915 Panama-Pacific International Exhibition in San Francisco. He sought to integrate science, technology, and the arts by "developing a core of materials with which the viewer can interact in order to explore the mechanisms of his own sensory perception."[53] It was a science center with a limited focus, no artifacts, and increased opportunities for individual experimentation and learning. As Oppenheimer later stated,

The Exploratorium provides exhibits, centering around the theme of perception, which are designed to be manipulated and appreciated at a variety of levels by both children and adults. The exhibits that explore and interact with the senses are fascinating in themselves; they also provide a genuine basis for interrelating science and art. Furthermore, understanding the mechanism of sensory perception leads to a flow of exhibits about beautiful and basic natural phenomena. By illustrating the reoccurrence of natural processes in a multiplicity of contexts, we convey a sense of unity among such processes and counteract any overly fragmented view of nature and culture.[54]

Two science centers came into being in the 1960s as direct offshoots of educational institutions: the Fernbank Science Center in Atlanta and the Lawrence Hall of Science in Berkeley. The former was started by a county school system, the latter by a university. Each has a rather special science education mission.

Fernbank Science Center grew out of a sixty-five-acre forest preserve, a school referendum, and the U.S. Office of Education's Elementary and Secondary Education Act and the National Defense Education Act grant programs. The DeKalb County Board of Education obtained a long-term lease to the forest in 1964 and funded the science center building through a school referendum in 1965. Operating and equipment funds came from the Office of Education.

Opened in 1967, the facility has become an integral part of the DeKalb County school system, with the forest and center serving as a nature and science laboratory. In addition to the forest, the science center has a planetarium, an observatory, an electron microscope laboratory, a meteorological laboratory, a library, and a majority of natural history exhibits.

Fernbank Science Center is dedicated "to bringing a greater understanding and appreciation of our natural world to mankind." More specifically, "Special programs, utilizing Fernbank's select staff and resources, bring enrichment to existing school curriculums, from kindergarten through graduate school, and provide quality in-service programs for teachers." In addition, "Fernbank activities are designed to offer the public an opportunity to increase their scientific awareness and enrich their leisure-time activities with enjoyable and constructive nature-related experiences."[55]

While Fernbank is nature-oriented, the Lawrence Hall of Science is primarily a center for research and development in science education. It was established by the regents of the University of California and named in honor of Ernest O. Lawrence, the university's first Nobel Prize winner, who was honored for his development of the cyclotron and other contributions to nuclear physics.

The Lawrence Hall of Science, which opened in 1968, is active in three broad areas: public programs and school activities, teacher training, and research and curriculum development. Unlike most science centers, its emphasis is inward rather than outward, serving as an incubator for improving science education materials and techniques.

The public programs and school activities include chemistry, physics, math, geology, biology, and astronomy exhibits; courses, field trips, and a summer science camp; and computer- and math-oriented activities in which more than fifty computer terminals are available for public use at the science center and about sixty-five remote ports are leased for instructional purposes by educational institutions in northern California.

In the teacher training area, Lawrence Hall of Science operates a mobile van equipped with teaching devices and science materials that travels to schools within a 200-mile area for teacher workshops; circulates materials to teachers from its science education library; helps teachers of deaf students to use computerized instruction; offers summer in-service teacher institutes; and directs teacher centers in outlying school districts.

As a research center in science education, Lawrence Hall designs, develops, and evaluates curricula and activities that promote science literacy at all educational levels. Among its research activities have been projects to improve outdoor biology instruction, better health understanding, science and mathematics instruction, the use of calculators, science activities for visually impaired students, the teaching of chemistry, and the use of computers and computer-controlled devices as instructional tools.

Worldwide Developments

While this explosion of new science and technology centers was taking place in the United States, significant developments were occurring in other parts of the world. Two large, multifaceted science centers, for example, were founded in the 1960s: the Evoluon in Eindhoven, The Netherlands, in 1966, and the Ontario Science Centre in Toronto in 1969.

The original idea for Evoluon came from J. F. Schouten, director of the Institute of Perception Research and scientific adviser to the Physical Research Laboratory in the N. V. Philips electronics company. But it was F. J. Philips, the head of the firm, who decided to enrich the Eindhoven community and the Philips organization with a permanent exhibition center on the occasion of the company's seventy-fifth anniversary. "The Evoluon presents an exhibition principally devoted to man and his relationship to technology in its relationship to man."[56]

The Evoluon is more than a corporate exhibition building. It is, in fact, a contemporary science and technology center in a European environment in many respects. Housed in a large and impressive 252-foot-diameter, mushroom-shaped building standing on twelve V-shaped columns, it contains three concentric ring-shaped exhibit floors, a children's gallery, an auditorium, a library, and a restaurant and makes extensive use of participatory and audiovisual exhibit techniques to further public understanding of science and technology.

Each of the three circular exhibit floors of the Evoluon has a different emphasis. The top floor deals with people and technology and shows the significance of industrial products for man and society in such areas as health, home life, recreation, education, communications, transportation, and industry. It also contains two balconies whose exhibits explain the importance of research in the physical and life sciences.

The middle floor focuses on technology in a more specific sense, with exhibits on vibration and sound, light and lighting, matter, controlled electrons, communication systems, production control devices, and computers. The lower floor, titled "Philips—an Industry," tells about the company's history, people, organization, and products.

An epilogue in the Evoluon's information pamphlet states that "It is not a museum; not a monument in stone commemorating geniuses of the past; not a collection of the products of a company, however fascinating such a collection might prove to be. The Evoluon is rather an industrial exhibition, a mirror of the past, present, and future, always seen through the eyes of man at grips with matter, wrestling with the design of his social system, winning many battles but never achieving the final victory."[57]

Some have said that the Ontario Science Centre in Toronto also is not a museum, but it most assuredly is a contemporary science and technology center. "During the last few decades," the center's fact sheet points out, "the image of the old-style museum—artifacts stored in glass cases like fossils inside blocks of ice—gradually has been thawed by the frustration of visitors kept at arms' length from displays. Demands arose for the development of exhibits allowing for direct visitor participation. With the opening of the Ontario Science Centre on September 27, 1969, this demand was met to the fullest."[58]

The science center contains nearly 550 exhibit units, most of which are designed to be operated directly by visitors. Large exhibit halls are devoted to physics, chemistry, earth science, biology, communications, transportation, general engineering, lasers, astronomy, and space, and smaller galleries to energy, environment, pollution, forestry, nuclear science, and the Canadian north country. In addition, there are eighteen small theaters, live demonstrations, a film series, and extensive educational and traveling exhibit programs.

"In the center's planning stage," explained J. Tuzo Wilson, director general of the facility, "some wanted a conventional historical museum to preserve examples of past Canadian technologies, but many of Canada's 800 museums, large and small, already did that. Fortunately, it was realized that here was an opportunity to educate the public about *today's* science and technology, upon which our lives and convenience so largely depend. It was finally agreed that about 20 percent of the exhibits should be historical to give a background for the 80 percent devoted to exposition of modern science."[59]

The Ontario Science Centre, which functions as an agency of the Province of Ontario government, is a split-level complex of three interconnected buildings with 450,000 square feet of floor space situated on a valley-and-knoll setting in a Toronto suburb. A primary aim of the center for youngsters "is to stimulate interest in science and technology through greater perceptual awareness. For persons of all ages, the Centre's concern is to show how the application of science and technology affects their lives and their environment—for better or for worse."[60]

In Japan, business and industry have taken the initiative in furthering public understanding of science and technology. The Japan

Science Foundation was formed in 1956 to conduct a variety of programs in science and technology, including making surveys and recommendations, cooperating with scientific organizations and universities, organizing courses and lectures, producing films and television programs, and supporting three contemporary science and technology museums in Nagoya, Osaka, and Tokyo.[61]

When the Nagoya Municipal Science Museum was opened in 1962, the city provided the building and the planetarium and the Japan Science Foundation financed the exhibits. The foundation established the Osaka Science and Technology Center in 1963. In addition to being a contemporary science museum, the center houses classrooms for courses, conference facilities, and the offices of scientific and technical professional societies.

The Science Museum in Tokyo, which opened in 1964, also was founded by the foundation. It is the largest of the science centers in Japan, covering some 236,000 square feet of space and featuring exhibits on electronics, polymers, nuclear energy, space exploration, transportation, electrical power, communications, architecture, natural resources, agriculture, machinery, appliances, and various fields of engineering. Approximately 450 of the newest machines and experimental apparatuses can be "touched or operated" by the visiting public.[62]

Among the other Japanese museums of science and technology are the long-established National Science Museum, operated by the Ministry of Education, Science, and Culture, which is expanding beyond natural history into technology, and numerous children's science centers that are supervised through the Board of Education.

Science and technology centers in developing nations have a somewhat different emphasis, as pointed out by Amalendu Bose, retired director of the National Council of Science Museums in India: "The level of understanding about science and technology by a vast majority of the people in these countries is insignificant if not nonexistent. Obscurantism and superstition still persist in many regions; fruits of science did not percolate down to the bottom of the social ladder. To many, science is no more than an object of awe and wonder. Naturally, the people, by and large, are not receptive to new ideas or adaptive to new techniques. But the state is committed to make them move forward, and to change the entire social superstructure."[63] He explained that it has become the obligation of the science and technology museum "to undertake the responsibility of educating the masses—literate, semiliterate, or even illiterate—about the social benefits of science and the need to imbibe a value and practice a way of life imbued with scientific outlook."[64]

In keeping with this philosophy, the government of India has established three science and technology centers and is planning a fourth. The first such facility was the Birla Industrial and Technological Museum in Calcutta in 1959, followed by the Visvesvaraya Industrial and Technological Museum in Bangalore in 1965 and the Nehru Science Centre in Bombay in 1978. A fourth center is being planned for New Delhi. All are administered and funded through the National Council of Science Museums.

The program at the Calcutta museum is typical of the Indian science centers. In addition to numerous contemporary exhibits and a few artifacts, the museum offers films, lectures, demonstrations, courses, and temporary exhibits; produces television programs; sponsors science fairs; circulates three mobile science vans; and operates three branch science centers in outlying areas.

One of the newest science and technology centers in a developing region is the Singapore Science Centre, which opened in 1977. In announcing plans for the government-operated facility, Minister for Science and Technology Toh Chin Chye stated: "The Science Centre was conceived to stimulate a continuing interest among children in the schools and the lay public in the principles of science and how they are used in technology. Establishing a science center may appear to be a paradox when in some industrialized countries there is a prevailing mood that science has not been able to solve the problems of their societies and the environment. In Singapore, however, it is not less but more knowledge of science and technology that we need, particularly so when the livelihood of the people of this city-state will increasingly depend upon our ability to develop new talent and skills to operate industries based on technological expertise."[65]

The Singapore Science Centre has contemporary exhibits on energy, communications, solar power, genetics, aerodynamics, population, nuclear science, space exploration, evolution, human birth, and automotive engineering in a lobby area and three galleries dealing with the physical sciences, life science, and special exhibits. About 40 percent of the 322 initial exhibit units are of a participatory nature. In addition, the center has a film program, several multimedia shows, science demonstrations, training courses, and a changing exhibit of industrial products.

R. S. Bhathal, director of the Singapore center, has described the increasingly important role of science centers:

A quiet revolution is now in progress as more and more people come to realize that the traditional museums can no longer be the preserve of a select coterie of people. They must be opened up to the hundreds and thousands of individuals who have suddenly found these places, which have themselves been undergoing changes, satisfying their need to obtain a non-formal and non-structured education. . . .

Science museums, by and large, have been the repositories of artifacts from the past and have tended not to worry too much about explaining science or making it understandable to the general public. Science centers, on the other hand, have taken it upon themselves not only to explain science to the public, but also to show how science is applied in industry. The very nature of their subject matter makes them innovative and experimental.

The emphasis in science centers is on contemporary science and its implications for society in general. Rather than maintaining a "hands-off" policy, science centers indeed encourage the public to touch exhibits, push buttons, turn cranks, and listen to taped messages on telephones so as to get visitors to participate in a learning experience.

Special education programs which take the form of discovery lessons, lecture-demonstrations, talks, science film shows, technology projects, and ecology study trips are normally run by the centers to supplement the formal offerings of local schools and colleges.[66]

Science and technology centers have evolved over several centuries, with the greatest refinement and growth taking place in the last three decades. They have developed from historic technical museums to exposition-inspired industrial museums to informal educational instruments of science and technology.

In both developed and underdeveloped nations, science and technology centers are popular, innovative, lively, and fast-growing. They also are constantly changing, experimenting, and improving their communication techniques to make contemporary science and technology more understandable and meaningful to millions of people each year.

3

Types of Science and Technology Centers

Contemporary science and technology centers are many things. No two are alike, yet all have a similar underlying philosophy that separates them from traditional museums and makes them effective vehicles of informal science education. The crux of that unifying philosophy is that learning about science, technology, industry, and medicine can be fun, and the most effective way to learn is through participatory exhibits of a contemporary nature. This process can occur at any one of three kinds of science and technology centers: comprehensive centers, specialized centers, and limited centers.

Comprehensive Centers

Comprehensive centers are the larger, broader, and more fully developed contemporary science and technology centers. Often they are the older and more extensive science and technology museums with large staffs, budgets, and attendance. Because of differences in origin, funding, and the nature of their exhibits and programs, comprehensive centers are generally categorized as industrially oriented, educationally oriented, or scientifically oriented institutions. Some science and technology centers, however, can span several categories.

Industrially Oriented Centers

An industrially oriented science center is a comprehensive museum that places considerable emphasis on industrial developments and relies to a large degree on exhibits and funding from industry. Some of these centers are technical and industrial museums, while others are relatively new institutions with an industrial approach.

In Europe, the Deutsches Museum in Munich, the Evoluon in Eindhoven, and the new National Museum of Science and Industry being developed in Paris would be considered industrially oriented centers. The German museum has numerous exhibits dealing with industrial technology, including products and processes, and its largely state-funded budget is supplemented by industrial support for specific exhibits. The Evoluon is owned and funded by the Dutch N. V. Philips electronics organization and is largely concerned with the impact of technology and the evolution of Philips's products and processes.

Five of America's largest and most popular science and technology museums—the Museum of Science and Industry in Chicago, the California Museum of Science and Industry in Los Angeles, the Franklin Institute Science Museum in Philadelphia, the Oregon Museum of Science and Industry in Portland, and the Center of Science and Industry in Columbus—can be categorized as industrially oriented, although nearly all could be considered educationally oriented and/or multifaceted centers. Each has numerous exhibits that point out the nature, development, and contributions of specific industries, such as the electrical, automotive, and chemical fields. They also receive substantial support from industry for exhibits and programs.

In Japan, industry provides exhibits and funds for the Nagoya Municipal Science Museum, Osaka Science and Technology Center, and Tokyo Science Museum. Although operated by the government and educational in nature, three science museums in India have numerous industrial exhibits and receive some support from industry.

Educationally Oriented Centers

Educationally oriented centers usually are operated by schools, universities, or governments, and present educational exhibits and programs that are primarily extensions of the classroom or laboratory.

The first educationally oriented center, as described earlier, was the Palais de la Découverte (Palace of Discovery) in Paris. At least three variations of education-based science centers can be found in the United States: the Lawrence Hall of Science, a branch of the University of California, Berkeley; the Fernbank Science Center in Atlanta; and San Francisco's Exploratorium.

In other parts of the world, educationally oriented science and technology centers generally are operated by the government and are closely tied to school offerings. The Ontario Science Centre in Toronto and the Singapore Science Centre, for examples, were built specifically to expand science education opportunities. Perhaps the best example of a nationwide system of educationally oriented science centers can be found in India, where the government operates the Birla Industrial and Technological Museum in Calcutta, the Visvesvaraya Industrial and Technological Museum in Bangalore, and Nehru Science Centre in Bombay, and is developing a fourth science center in New Delhi. There is, however, at least one educationally oriented center that is operated by industry—the Alfa Cultural Center in Monterrey, Mexico, is sponsored by the Alfa Industrial Group.

Scientifically Oriented Centers

A third form of comprehensive science and technology centers encompasses natural history and/or other fields in addition to the physical and life sciences. The prime example of this type of science center is the Museum of Science in Boston, which started as a natural history museum, then expanded to include exhibits and programs in the physical and life sciences. The Cranbrook Institute of Science in Bloomfield Hills, Michigan, the Science Museum of Minnesota in St. Paul, and the Maryland Science Center in Baltimore emerged from similar backgrounds.

Specialized Centers

While comprehensive science and technology centers tend to be broad based in their content and approach, specialized science centers focus on a more narrow aspect of science and technology, such as health, energy, transportation, space, and nature, and generally are smaller than their comprehensive counterparts. They basically are contemporary in content, participatory in approach, and educational in philosophy.

Health Centers

Health museums are among the most numerous specialized science centers. They are concerned primarily with health education and usually have contemporary medical and health exhibits, health-related courses, and other activities to improve the overall health of the community.

One of the first science centers in the health field was the Cleveland Health Education Museum. It has exhibits on the brain, nervous system, sight and sound, human development, birth process, blood vessels, body functions, microbes, and other health areas. It also offers lecture-demonstrations, films, courses, and community service programs in the health field.

The Dallas Health and Science Museum and the Robert Crown Center for Health Education in Hinsdale, Illinois, present two other approaches. In addition to its health exhibits and programs, the Dallas museum has a planetarium, a radio station, and exhibits on communications, earth sciences, space, and veterinary medicine. The Crown Center minimizes the museum visitation function and emphasizes the use of exhibits and audiovisual and teaching aids in courses on general health education, drug abuse preventive education, life begins/family living education, and environmental education/human ecology.

Energy Centers

A number of science centers devoted largely to furthering public understanding of the sources, generation, and use of energy have been started in recent years by government agencies. The American Museum of Science and Energy in Oak Ridge, Tennessee, is the largest and best known of these energy-oriented institutions. It is operated by the Oak Ridge Associated Universities for the U.S. Department of Energy in a community noted for its scientific research and development in the nuclear field. Once confined to nuclear exhibits and programs, the Oak Ridge museum now covers all fields of energy and operates an extensive traveling van exhibit program.

In Mexico City, the Federal Commission of Electricity, which is charged with generating, transmitting, and distributing electricity in Mexico, operates the Museo Tecnológico (Technological Museum). The museum originally dealt exclusively with electrical principles and systems but now also includes exhibits on transportation, space, and other fields.

Transportation Centers

Automobile, railroad, aircraft, and marine museums are common, but science centers specializing in transportation are relatively rare. Most transportation museums concentrate on one mode of transportation and usually are more concerned with collecting and preserving artifacts than with communicating scientific and technical information to the public. Such museums become science centers when their exhibits and programs broaden to include the scientific and technical aspects of transportation.

The Verkehrshaus der Schweiz (Swiss Transport Museum) in Lucerne mixes the old with the new and adds a scattering of other exhibits in communications, astronomy, and space exploration. In addition to its extensive collections of automobiles, locomotives, airplanes, ships, and related communications equipment, it also has a planetarium and features a popular multimedia show tracing American and Soviet exploits in space.

Another transportation museum that has broadened its offerings is the Museum of Transport and Technology in Auckland, New Zealand. Although most of the exhibits are historic in nature, the museum has sought to incorporate contemporary techniques, add technological exhibits, and expand its activities for the benefit of the general public.

Space Centers

With the space age has come museums of space artifacts, exhibits, and programs. These space science centers feature participatory exhibits and programs about the past, present, and future of space exploration as well as relics and replicas relating to the space effort. Sometimes they operate planetariums and have astronomical exhibits. The Alabama Space and Rocket Center in Huntsville is representative of space science centers. Operated by the State of Alabama near the federal government's space and rocket facilities in Huntsville, the center has a collection of early missiles and rockets, actual spacecrafts, and other space equipment, as well as informational, participatory exhibits. A similar but smaller facility is the Michigan Space Center in Jackson.

The new Hong Kong Space Museum has a planetarium as well as astronomical and space artifacts and exhibits. The Smithsonian Institution's National Air and Space Museum in Washington also falls in the space center category, although it includes aviation. Both museums skillfully blend historic artifacts with contemporary messages in depicting developments in space. Interactive and audiovisual techniques are utilized in many exhibits to involve and inform museum visitors.

Nature Centers

Nature centers usually are neither natural history museums nor full science and technology centers but hybrids that are considered science centers because they deal with the study of ecology and an appreciation of trees, flowers, and wildlife.

The Charlotte Nature Museum in North Carolina has been one of the pioneers in the field. In over three decades it has grown from basically a nature walk and study museum to a broader-based science center with exhibits and programs on health, perception, astronomy, local history, and ecology. The museum also offers courses, operates a traveling exhibit van, and supplements school programs, and a new downtown satellite facility called Discovery Place makes it possible to broaden its science and technology exhibits and educational programming.

Two other North Carolina centers that have expanded the scope of their operations are the Nature/Science Park in Winston-Salem and the North Carolina Museum of Life and Science in Durham. The Peninsula Nature and Science Center in Newport News, Virginia, is doing likewise. In West Hartford, Connecticut, Environmental Centers Inc. operates a nature center as well as a planetarium and an aquarium. Atlanta's Fernbank Science Center, now considered a comprehensive science center, began as a nature center.

Limited Centers

Limited science centers are either small science and technology centers with few offerings or other types of museums with science center components, such as children's museums, natural history museums, history and science museums, and multipurpose museums with a portion of their exhibits and programs devoted to contemporary science and technology.

Small Science Centers

Philosophically, small science centers are similar to comprehensive centers but are not able to offer as extensive a range of exhibits and programs nor serve as many people because of their limited facilities, staffs, and/or budgets. Among the small centers are the Des Moines Center of Science and Industry, Science Museum of Virginia in Richmond, Detroit Science Center, Reuben H. Fleet Space Theater and Science Center in San Diego, Impression Five Science Museum in Lansing, Michigan, Omniplex in Oklahoma City, and John Young Museum and Planetarium in Orlando, Florida.

Many of these science and technology centers are relatively new but are growing rapidly. As they expand, they may become comprehensive centers. Others will remain small because the community does not need a major facility or because sufficient interest and support cannot be generated.

Museums with Science Center Components

An increasing number of traditional museums are incorporating science and technology center concepts and exhibits into their programs. Other institutions, such as children's museums, have expanded the scope of their activities to include scientific and technological elements.

Participatory techniques are not new to children's museums, but substantive exhibits and programs that deal with scientific principles, technological applications, and health considerations are a recent development. The Brooklyn, Indianapolis, and Boston children's museums have included sections on the physical and/or life sciences in their newly opened facilities.

Some natural history and history and science museums are moving in the same direction. They include the National Science Museum in Seoul, Korea; the Museum of Science and Space Transit Planetarium in Miami; Museum of Science and Natural History in St. Louis; National Science Museum in Tokyo; Kansas City Museum of History and Science; Museum of Natural History and Science in Louisville; Springfield Science Museum in Massachusetts; and Milwaukee Public Museum.

General-purpose museums sometimes have participatory exhibits and programs dealing with science, technology, industry, and/or health, such as the Museum of Arts, Sciences, and Industry in

Bridgeport, Connecticut; Evansville Museum of Arts and Sciences in Indiana; and Lakeview Museum of Arts and Sciences in Peoria, Illinois.

The Future

The science and technology center movement continues to evolve and grow. Definitive statements are not possible, but there are some indications for the future.

The number of science centers will accelerate in both developed and developing countries. Although they probably will take different routes, it is clear that the emphasis will be on educationally oriented centers and small science centers.

The cost of developing a comprehensive science and technology center has become almost prohibitive. With the decline of personal fortunes and rise of corporate restrictions, it is extremely difficult to build a major new center without government support. National governments, however, are becoming increasingly interested in science centers as instruments for social uplifting and furthering public understanding of science and technology.

The new contemporary science and technology museums in India, Thailand, and Singapore are likely to be followed by other educationally oriented science centers funded by governments in developing countries. Plans already are being discussed in Chile, Malaysia, Ghana, and the Philippines. France and Spain are among the developed countries formulating ideas for major science centers. It also is likely that the traditional technical museums of Europe will include more contemporary materials and techniques in their exhibits.

In the United States, it appears that the proliferation of small and medium-size science and technology centers of an educational nature will continue. Having seen the success of science centers in large urban areas, community leaders and groups will want similar facilities tailored to the needs of their regions. This trend probably will be accompanied by a broadening of specialized science centers and increased contemporary and participatory science exhibits at other types of museums.

In general, science and technology centers will play a more important role in informal science education throughout the world. The emphasis will shift from a few large comprehensive centers to a network of small and medium-size science centers. In the process, government will assume a greater share of funding these institutions.

These trends probably will have little impact on the funding of science center exhibits and programs by industrial firms. Science centers will continue to seek sponsors for exhibits, but they will exert greater control over such exhibits to make certain that they are objective presentations.

At the same time it is likely that science and technology centers will assume increased responsibility for public education on science policy issues and become more responsive to environmental and other societal implications of science and technology.

A planning report for a new museum of science and technology in Hong Kong pointed out the significance of such science centers:

Science is constantly producing new facts and theories, and technological advances are rapid. A science museum is a place where, throughout their lives, people can have direct contact with new technology and can experience the basis for scientific processes. Science instruction for the general public is increasingly important because citizens cannot participate in, and understand, political and social events of our time without a basic literacy in science and technology. Perhaps even more important, literacy in science is important for the quality of life. Understanding the living and physical world around us leads to a fuller life and is as important as an understanding of art and literature.[1]

4

Starting a Science Center

Birth of an Idea

A science and technology center is spawned in many ways. A vacation or business trip to another city with a science center may ignite a desire to have a similar facility. The Junior League may recognize the community need for a science center and undertake it as a project. A science center may be looked upon as a worthy memorial by a wealthy individual or family that is willing to provide the startup costs. A parents and/or teachers group may feel the need for an informal education center in science. The local government or chamber of commerce may believe a science center will improve the community's image and attract tourists or industry. A national government or university may be convinced that a science center is an excellent way to improve educational opportunities, advance social uplifting, and further public understanding of science and technology.

These and other circumstances have been influential in starting science and technology centers in both developed and developing nations. It does not take much to plant the idea. The difficulties occur in implementing plans for a science and technology center. Too often the individuals and groups that initiate the science center idea plunge into the project without adequate preparation, study, or consideration of all the factors involved in such a decision. The result is that the science center is doomed to failure, plagued by problems, or does not reach its potential.

One of the most frequent difficulties is the desire to duplicate a science and technology center located elsewhere rather than trying to tailor the center to local needs and resources. Another common failing is the launching of a science center without sufficient community involvement and financial support. It is one thing to open a science center and another matter to keep it in operation.

Mahmoud Mesallam, founding director of the Museum of Science and Industry in Cairo, has told of his difficulties after the museum was established by the Egyptian government in 1969 in one of the palaces of the former King Farouk. Despite considerable planning, outside consulting, and government support, the museum failed to attract the public. After 2,000 visited the museum on the first day, the attendance dropped to only 300 per month. In addition to an inappropriate site, Mesallam cited other factors responsible for its lack of success:

1. The exhibits were built to interest those who designed them and did not meet the needs of the different sectors of the public.
2. The models were too big for the ideas they had to illustrate.
3. There was no sequence. The exhibits were put in the positions which best suited them, and not according to any logical sequence.
4. The exhibits were silent and could not arouse the interest of the visitors.
5. There were too many words to explain each exhibit. Visitors do not want to come to museums to read books . . . .
6. The exhibits were much too crowded.
7. There should have been demonstrations of big national projects.
8. There were no exhibits to show the recent scientific and technological achievements in developed countries.[1]

In other words, the museum was uninspiring, failed to communicate, and overlooked contemporary science and technology. Mesallam was fortunate to be able to correct many of the problems by rearranging the exhibits, adding exhibits on Egypt's major technological projects, scheduling traveling science exhibits from abroad, installing participatory devices, organizing a science club program, initiating summer camps, and developing an outreach exhibit and film program to villages to stimulate interest and communicate more effectively. As a result, the museum's attendance increased from 300 a month to 2,000 a day.[2] But the museum later faltered from the lack of government support.

A number of helpful guides have been published to assist in starting and operating museums, including science and technology centers. Among the most useful are *So You Want A Good Museum, The Organization of Museums, Basic Museum Management, Thoughts on Museum Planning,* and *Starting a Science Center?* (see selected bibliography).

This chapter discusses some of the principal areas of concern in starting a science and technology center and seeks to point out some of the necessities, choices, and pitfalls. It is not intended to answer every question but rather to stimulate creative thinking in planning a science center.

Founding

Starting a science and technology center can be a somewhat difficult process. It involves a number of crucial steps. For example, it is necessary to decide what type of science center it should be, who it will serve, what its legal status should be, where it should be located, what type of building it should be housed in, how it will be supported, who will govern it, and how it will be staffed. Answers to such questions need to be found *before* the science center begins operating.

Planning Committee

Few people have the power and the resources to start a science and technology center alone. Thus, one of the first steps is to involve others—friends, neighbors, parents, teachers, businessmen, scientists, engineers, school administrators, government officials—who are interested in the idea. It may be appropriate to form a committee that will explore and spearhead the organizing effort. The committee members should be able to contribute time, ideas, influence and/or funds for the project.

In addition to discussing the possibilities of a science and technology center, the committee members should seek to determine whether the community really needs and wants a science center. This can be done through interviews, open meetings, and surveys.

This committee—or a smaller task force—should obtain information about science and technology centers in other communities and countries. The fact-finding group certainly should contact the Association of Science-Technology Centers (ASTC) and possibly other appropriate museum organizations, such as the International Council of Museum Committee of Museums of Science and Technology, American Association of Museums, Association of Science Museum Directors, American Association of Youth Museums, and/or American Association of State and Local History. If possible, it also would be desirable for representatives of the planning group to visit three or four typical science and technology centers. The visits could be most helpful in making basic decisions about the nature, structure, programming, and funding of the proposed facility.

At this point, it may be possible for the committee to decide on whether the science center idea should be implemented, modified, postponed, or abandoned. Sometimes the planners will seek professional assistance before making such a decision, but more often the investment in museum consultants, designers, and/or planners takes place after the commitment is made to proceed with the project.

Professional Advice

Professional assistance can be invaluable in planning, designing, promoting, and raising funds for a new science and technology center. Various types of expertise are available for a fee.

As the planning group prepares to implement the proposed science center, it frequently is helpful to obtain the guidance of a recognized authority in the field or a panel of three or so experienced science center directors. Such outside expert advice usually is provided on a one- or two-visit basis.

Sometimes this step is bypassed and a museum planner, exhibit designer, or architect is hired to assist in developing the science center concept, designing the exhibits, and/or making plans for a new building or renovating an existing structure. It usually is preferable to use a professional museum planning or exhibit design firm—some groups offer both services—although an architectural firm can be brought in later as building plans develop.

The objective of such professional assistance at this stage should be to develop a workable science center concept that can be presented to the community for support. The professional planner-designer should be able to suggest one or more suitable locations; offer guidance on an old or new building; develop exhibit, program, and operational approaches; and provide rough cost estimates. The end result may be a report, sketches, a model, and/or a slide presentation, depending upon the committee's needs and budget.

With such information and materials, it should be possible for the committee to proceed with planning, promoting, organizing, staffing, and fund raising.

If the committee should decide to launch a major effort to raise funds for the project, it may wish to hire another type of professional: a fund-raising firm. Some companies merely offer counseling and/or help determine the extent of community support while others also organize and direct the fund-raising campaign.

Two other types of professionals will be useful as the institutional plans evolve. An attorney should be used in drawing up the bylaws and filing for incorporation and nonprofit tax status. An accountant can be helpful in setting up the financial accounting.

Type and Purpose of Center

In planning a science and technology center, a founding group must decide early on the nature and objectives of the center. A science center can be industrially, educationally, or scientifically oriented; it can emphasize an area such as health, energy, transportation, space, or nature; or it can serve only a partial science center function because of its size, scope, or responsibilities.

The orientation of a science center is important because it influences the content, budget, and approach of the facility. It helps determine whether the emphasis will be on collections or exhibits, whether curators are hired, and whether most funds will go for public programs and influences the location, type of building, amount of funding, and whether it should have such facilities as a planetarium, classrooms, or nature trail.

In making such decisions, it is necessary for the planning committee to examine what museums and other facilities already exist in the community, region, or nation. The committee also should seek to determine what type of science center would be most useful and receive the greatest support. The nature and size of the community also will affect the final decision.

Every science center should have stated objectives. It is extremely helpful to set such goals in the formative stages. Adoption of a list of objectives can come before or after the decision on the type of science center, but they should be an integral part of the planning process. They serve as a road map in planning and operating the museum.

Legal Structure

Every science and technology center has a legal status. Therefore, it is necessary to give consideration to such legal matters as the institutional constitution, bylaws, incorporation, and nonprofit tax exemption during the formative stages.

The constitution and bylaws provide the basis for incorporation, tax exemption, and legal responsibilities for operating a science center. They generally describe the purpose of the institution and provide for a governing board, officers, director, committees, members, financial matters, and procedures for changing the bylaws.

Incorporation makes it possible for a science center to enjoy permanent status, hold property, enter into contracts, accept contributions, and limit the liability of its board and members. Incorporation applications usually are filed with the secretary of state in the state in which the institution is located.

Another vital legal step is to obtain tax-exempt status as a nonprofit organization. In the United States, application is made to the Internal Revenue Service under provision 501(c)3 of the Internal Revenue Code. Approval of the application exempts an organization from payment of taxes and enables it to receive tax-deductible contributions.

A lawyer should prepare such legal documents. Frequently, an attorney who is a member of the planning committee or governing board may contribute these services.

Site Selection

Selecting the location for a science and technology center requires consideration of all possible options. Ideally, the decision should be based on the job the science center seeks to perform. Unfortunately, other factors, such as cost, the availability of land, and pressures to use existing structures, usually influence the decision.

Whenever possible, site selection should consider an institution's functional and operating needs, as well as costs and available buildings. In general, central locations are preferable. But it may be more logical, for instance, to locate a science center with an emphasis on nature studies in a park or outlying area of a city.

A site also should be accessible to the market served, offer adequate parking, provide expansion space, and be compatible with the neighborhood environment.

Building

Most founding groups would prefer to build a new structure to house a science and technology center. In this way, at least theoretically, the building could be designed to accommodate all the needs of the center. However, it frequently is not possible to fund the construction of a new building. As a result, it becomes necessary to adapt an old building for the science center.

The type of building—whether old or new—has considerable impact on the success of the institution. The planning committee should consider such factors as the building configuration, floor loads, door openings, ceiling heights, elevators, electrical power, exhibit space, storage, office space, restrooms, sales areas, classrooms, heating and cooling systems, receiving dock, parking, sprinkler systems, and maintenance and energy costs.

Among the common existing buildings suggested for science centers are abandoned railway stations, old courthouses, fair buildings, unused mansions, and vacant factories, warehouses, and store fronts. All of these are possibilities, but the cost of converting, maintaining, and operating these old structures must be compared against the cost of a new building.

Depending upon a science center's needs, a rehabilitated building may be the best investment. On the other hand, if funds permit, construction of a building tailored for a science center may be a better decision. Sometimes it is a two-step process, with the science center being launched in a converted building while funds are sought for a new structure. But such a tactic can backfire if the initial operations are a disappointment to potential donors.

An architect should be brought into the planning early whether restoring or constructing a building. For the most effective utilization, the specific needs of a science center should be spelled out by the planning group. Most architects have not designed a science center or museum before and need as much input as possible from the founders, the exhibit designer, and experienced science center administrators.

Funding

There are two phases to funding science and technology centers: the startup money for housing and equipping the center and the operating funds to staff and run the facility. A founding group has to be concerned with both aspects.

Science centers are expensive and it frequently takes a major effort to raise the funds to start them and even more to keep them operating. A community need or good concept is not enough to raise the necessary funds. It takes excellent leadership, good organization, extraordinary effort, and a solid funding formula. It even may require the hiring of a fund-raising firm to supplement the efforts of the founding group, governing board, volunteers, and/or staff members.

In the past, philanthropists frequently provided the funds to start science and technology centers in the United States. But they are a vanishing breed in today's society, and it nearly always is necessary to solicit contributions from individuals, companies, and foundations; to hold fund-raising benefits like dinners and auctions; and to attempt to obtain startup funds from local governments. In most other countries, the funding of science centers usually comes from the cultural, educational, scientific, or industrial ministries of the national government. However, an increasing number of government-operated science museums are obtaining funds from industry and other private sources for exhibits and other activities.

An operating budget should be part of the startup planning. Science centers are not profit-making enterprises; thus it is necessary to supplement earned income from admissions, memberships, sales, rentals, and/or fees. Ideally, an endowment or investment fund—created with earmarked gifts—can make up the difference with dividends and interest. But most museums are not that fortunate.

Sometimes it is possible to obtain continuing support from city or county government, the school system, or a special taxing authority. In cases where public funds are used to support a science center, it may become necessary for the institution to be part of the government body rather than to operate as a private organization.

More often, the operating deficit has to be made up by contributions, grants, and fund-raising events.

Governing Board

When starting a science and technology center, the founding group should give consideration to the composition of the governing body—usually known as the board of trustees—that will be legally responsible for operating the institution. It would be a mistake merely to convert the planning committee into the governing board, since the founders may not be the best choices for such a difficult job.

A board of trustees sets policies, controls finances, selects the principal administrators, approves major decisions, and sees that the purposes of the institution are carried out. It usually helps to raise funds when needed. Because of these legal, financial, and leadership responsibilities, every effort should be made to select the best qualified people for the board. They usually are civic leaders, educators, scientists, businessmen, city officials, and representatives of various segments of the community.

The governing board normally consists of from twenty to forty members, although some have as few as six or as many as fifty. Board members generally are elected for staggered terms and meet every two or three months. Most boards elect officers annually and have an executive committee to make decisions between meetings.

The key to the effectiveness of a governing board is the chairman and/or the executive committee. The chairman must take the initiative on policy decisions, major appointments, and fund raising and must have stature and leadership abilities to command support on the board and in the community. At the same time, the chairman and board should not meddle in the day-to-day operations of the science center, which is the responsibility of the director and his or her staff.

Staffing

The appointment of a director may take place before or after a science and technology center becomes a reality, but the remainder of the staff usually is not chosen until the plans are well developed and the funds are available to proceed with the project. The governing board selects the director and frequently approves the appointment of other key staff members, such as department heads.

The staffing of a science center can be quite different from that of a traditional museum, especially if the institution is not collection oriented. In many instances, collections either do not exist or are minimized because of the emphasis on participatory exhibits and educational programs. As a result many science centers do not have curators, or their role is quite different from what it is in a museum with many "hands-off" objects of intrinsic value.

Science center directors come from many backgrounds, depending upon the needs of the institution. A person with experience in the museum field, perhaps as an assistant director, curator, or education director, frequently is preferred. Other institutions seek to recruit able administrators with successful managerial, educational, scientific, communication, and/or fund-raising background. A science center director must be able to communicate, motivate, and work with people; organize and implement successful exhibits and programs; and obtain the necessary financial, promotional, and public support to operate the facility.

Programs

The range of programs at science and technology centers varies with the nature, size, and budget of the institution. Virtually all science centers, however, have exhibit and education programs, and many have collections, membership programs, and community services. Some also conduct research and engage in other programming activities.

Collections

Collections of objects of intrinsic value are the heart of traditional museums, but they frequently are incidental to the main thrust of a science and technology center. In fact, some science centers do not have any collections in the usual sense.

One of the most significant decisions to be made by a founding group in planning a new science center is whether to have or to emphasize collections of artifacts. Most comprehensive science and technology centers do not have extensive collections or curators to gather, conserve, preserve, study, and interpret the objects. But collections do play an important role at many specialized and limited science centers.

With collections, it is necessary to provide for staff, work space, record keeping, and storage, as well as exhibits and educational programming, in the planning and budget. It also is essential to be selective in acquiring collections because of the cost of restoring, storing, and protecting delicate, valuable, or large objects. A science center never should agree to keep or display any object forever. It must have the flexibility to remove from display and dispose of any object or exhibit at any time.

Exhibits

The success or failure of a science and technology center usually is determined by the quality of its exhibit program. The effectiveness and popularity of the exhibits in communicating scientific and technological information to the visiting public have a great bearing on the interest in and support of the institution.

The key to exhibits at science centers is participation—the opportunity to touch, manipulate, and interact with exhibits—and the institution's ability to maintain the exhibits. Participation can take many forms, such as touching objects, using telephones for narrations, operating working models, showing films or slides, and walking through the woods.

Because science and technology are forbidding to many people, it is necessary to have exhibits that are enticing, enjoyable, and understandable. It is a difficult assignment but essential if the exhibits are to be productive learning experiences.

The "see-but-don't-touch" approach may be necessary when valuable collections are involved, but even such historic exhibits can be made participatory with supplemental audiovisual materials and question-and-answer devices. Some collection-oriented and historical exhibits may be arranged sequentially. Studies also have shown that the public does not tour a museum in an orderly fashion and that visitors flow randomly from one exhibit unit to another. People tend to go to exhibits that appeal to them personally or that attract their attention with movement, colors, size, or sounds.

Regardless of a museum's nature, size, or support, it will not be possible to cover all fields of the physical and life sciences, engineering, medicine, technology, industry, and natural history. Therefore, it will be necessary for the planning committee, the governing board, and/or director and staff to make choices and select priorities for the exhibit program.

Most science centers have one or more exhibit designers on their staff and a number of craftsmen to keep the exhibits operating. Some science centers design and build all of their exhibits; other centers contract with outside firms for designing and fabricating exhibits and merely handle the planning and maintenance with their own staff. More often, however, it is a combination of inside and outside efforts.

Education

Although almost all science center activities are educational, the "education program" usually refers to specific activities organized for educational purposes beyond the exhibits and routine operations. They may range from presenting scientific demonstrations for school groups to offering lecture tours, courses, workshops, seminars, films, science kits, school loan programs, field trips, publications, teacher institutes, outreach programs, nature trails, and summer camps.

Many science centers have an education department to conceive, organize, and conduct the educational activities. Part-time teachers, volunteers, and docents frequently are used to supplement the education staff, particularly in presenting scientific demonstrations, instructional courses, and school programs at the science center as well as in the schools.

Teacher guidebooks, quiz sheets, films, and publications for supplemental study frequently are provided to schools by science centers as part of their educational services. For school group visits to be meaningful, it takes careful planning by both staff members and teachers.

Research

Scientific research to further knowledge in a particular field is rare among science and technology centers, except for those that have natural history or other extensive collections. The research usually is conducted by curators or staff members with research responsibilities.

Many science centers, however, do research to gather information for exhibits and/or to evaluate exhibits and programs. Such research normally is carried out by exhibit and education staff members and is aimed primarily at improving the content and effectiveness of exhibits and programs.

Membership

A membership program usually is initiated to do a better job of serving interested parties and to develop continuing support for an institution. Most membership programs have a graduated dues schedule for students, adult individuals, families, corporations, and various supporting memberships, including a life membership category. All the memberships must be renewed annually except the life memberships.

Among the typical benefits of membership are free admission, invitations to evening open houses and exhibit openings, discounts on store purchases and educational programs, and subscriptions to the museum newsletter, calendar, and/or journal. Some science centers also have a members' lounge and organize special tours for members.

In addition to receiving membership dues, a science center hopes to develop an annual giving and bequest program, obtain volunteers and docents, and convert the members to goodwill ambassadors for the institution. Membership support can be most helpful, but it also can cost more to obtain and service members than the funds received unless well conceived and administered.

Community Services

Most science and technology centers offer a variety of community services, some of which relate to an institution's science education mission and others that are provided as a convenience to the community. A science center must be careful, however, that it does not stray too far from its basic purposes. Dilution of activities could affect an institution's community role, support, and even tax-exempt status.

Science centers frequently serve as showcases for science fairs, industrial education programs, and various types of science-oriented hobby groups. Increasingly, they also are an outlet for other types of community events, such as arts and crafts exhibitions, ethnic and cultural activities, and minority programs. Some activities are designed to acquaint the underprivileged, aged, or handicapped with science, while others are primarily to service the community.

Administration

The administration of a science and technology center includes general management and planning, recordkeeping, building maintenance, security, store and food sales, fund raising, public relations, budgeting and financial reporting, and other such functions. A science center needs a supporting staff to provide the overall leadership, care, protection, funds, promotion, and fiscal management necessary to operate the institution.

Management and Planning

Sound management is essential to any science and technology center, but it does not come easily. It requires an effective organizational plan, a qualified director and staff, financial planning and controls, efficient administrative services, and a lot of hard work.

The director is the chief administrator, planner, and fund raiser of a science center. He or she is the bridge between the board of trustees and the staff, the principal organizer and implementer of the institution's goals and aspirations, and the person responsible for seeing that the science center operates smoothly, effectively, and within its resources.

A science center's organizational pattern usually is developed by the director in consultation with the board of trustees. In addition to determining the best organizational structure, it involves decisions on the assignment of responsibility, the level of funding, and the selection of people.

Most science centers have a director and possibly an assistant director; a number of department heads, usually for exhibits, education, business affairs, building maintenance, and public relations; and perhaps separate offices for collections, security, development, membership, and volunteers (although they sometimes are included within major departments).

The responsibilities of the various departments and offices as well as institutional procedures should be put in writing. It also is desirable to have job descriptions and pay scales for administrative purposes and to schedule regular staff meetings to discuss plans, problems, and needs.

Business Affairs

A knowledgeable business office or department is a necessary ingredient of any science center operation, whether it consists of a single bookkeeper or a staff headed by an accountant or a business school graduate. This department generally is responsible for financial records, payroll, budgeting, fiscal control, sales activities, and financial planning. In some cases, it also is entrusted with security, building maintenance, fund raising, and/or membership.

Accounting and financial reporting are the foundation of business office activities. Some centers operate on a cash basis of accounting, but the preferred method is to prepare financial statements in accordance with "generally accepted accounting principles," as prescribed by the American Institute of Certified Public Accountants.

Building Maintenance

Whether a science center rents or owns its building, it is necessary to have one or more persons responsible for cleaning, heating, cooling, repairing, and caring for the building. The size of the structure and the nature of the operations usually determine how many and what types of maintenance people are needed.

The "building engineer" will range from a general maintenance person to a professional engineer, depending upon the requirements. He or she also may be charged with capital improvements, although this responsibility usually is shared with the business officer and an outside architect. A building department normally operates independently or as part of the business office.

Security

Every science and technology center should have a security system, regardless of whether it has valuable objects on display or no collections. In some instances, security may be the responsibility of a staff member with other duties. Other times, an institution may have a full-time security chief who usually reports to the business officer. The security force ranges from guides and volunteers to uniformed guards and city police. Science centers with large attendance sometimes have one or more city police officers assigned to supplement the institutional force.

Public Relations

Every science center has an obligation to let the public know about its activities. It also needs publicity and publications to attract people to use its services and to provide funds for its operations. For these reasons, most science centers have one or more persons in a public relations or public information department.

The public relations function calls for communicating an institution's objectives, programs, and needs to the trustees, members, employees, contributors, volunteers, schools, taxpayers, government officials, neighbors, tourists, media, and general public.

Fund Raising

The raising of funds to support the operation of a science and technology center is a continuing need. It is the basic responsibility of a director. However, many science centers have established a development office or department to assist in obtaining funds from individuals, companies, foundations, government agencies, and other sources.

It usually is the responsibility of the development office to maintain the contribution and grant records, develop prospect lists, prepare fund-raising materials, handle the solicitation mailings, make contribution presentations, submit grant applications, plan fund-raising events, acknowledge contributions, and organize and coordinate the fund-raising efforts of staff members, trustees, volunteers, and others.

Fund raising is a difficult and frustrating job, but it is imperative to the survival of the institution. A development officer can make the task much easier for the science center and the director by helping to identify likely prospects and by preparing the necessary tools to pursue them.

II

Organization and Administration

5

Institutional Objectives

What Is the Purpose?

Every science and technology center has a purpose, but too often a founding or governing body mistakenly assumes that it is obvious or generally understood and that there is no need to have a statement of objectives. A clearly defined concept that includes a list of specific objectives, however, can be invaluable in starting and operating a science center. Such a road map can help guide the founders, the governing board, the staff, and others in building, programming, funding, hiring, planning, and maintaining an institution.

Typically, institutional objectives are part of the original concept, constitution, or bylaws; however, they sometimes are adopted by the board or staff at a later date.

Traditionally, museums have sought to collect, preserve, and interpret objects of intrinsic value. But the advent of science and technology centers and other developments have changed this basic mission in many instances. Jacques Barzun, the noted Columbia University humanist, has attempted to define a museum's purpose with these developments in mind: "to convey, preferably in some organized form, knowledge with pleasure—aesthetic pleasure sometimes, and the simplest pleasure of knowledge at other times."[1]

Whatever the pleasure, it must be defined. As Ger van Wengen of The Netherlands has pointed out: "There has been a distinct shift in recent times from the strongly object-oriented interest of museums to a new public-oriented interest. This is a distinct shift in emphasis, but one which is extremely important for the current development of museums. Such development calls for a well-defined museum policy, one which can be given concrete form by defining appropriate aims."[2]

A conference on the purpose, financing, and governance of museums emphasized that "Museums cannot merely drift through a changed economy, a changed society, and an expanded and rapid transportation and communication system. At the very least, a statement of purpose and objectives is an essential management tool."[3]

In general, science and technology centers seek to further public understanding and appreciation of science and technology. However, within this broad purpose, they usually have more specific objectives. Among the common objectives of science centers are to preserve and interpret the nation's or region's scientific and cultural heritage; to explain scientific principles, technological applications, and social consequences; to provide first-hand experience with science; to supplement school programs in science; and to help produce an informed citizenry to make more intelligent public policy decisions on scientific, technological, and industrial issues.

Some institutions are even more specific in stating their objectives, focusing on such local needs as environmental field trips; preschool classes in science; student science fairs; health education programs; outreach to senior citizens, the handicapped, and/or minorities; summer camp programs; and an examination of the future from the standpoint of science and technology.

One of the outcomes of the 1978 International Meeting on the Planning of Museums of Science and Technology in Developing Countries in Manila was a list of objectives for science and technology museums:

1. To show the indispensability of science and technology.
2. To stimulate interest in science and education so that, in the future, more and more people may be attracted to science, engineering, and research.
3. To show the application of science and technology in industry and for human welfare.
4. To encourage creative talent in the younger generation.
5. To disseminate scientific and technological knowledge among citizens of different ages and educational levels.

6. To demonstrate the importance of technical achievements along with the advancement of science and technology.[4]

The meeting report also stated that "The natural and technical sciences have no national boundaries. The laws of physics and those of telecommunications are the same in both rapidly developing and highly industrialized countries. With this in mind, it would seem that museums of science and technology and science centres throughout the world could adopt a universal form and speak the same language."[5]

Unfortunately, although the subject matter may be similar, science museums and centers differ from country to country and even from city to city. It is because the circumstances and objectives that give rise to these institutions frequently are quite different from one place to another. It is unrealistic to believe they should follow "a universal form," or even to "speak the same language."

As stated by Theo Stillger, general director of the Deutsches Museum in Munich, "The objectives of a museum of science and technology are as different as the needs of the target groups for which they were intended. Each museum has its own history which it cannot and must not deny; each country and each educational system raises different questions for which the museum of science and technology must provide an answer."[6]

Developing a List of Objectives

Institutional objectives take many forms. The Center of Science and Industry in Columbus, Ohio, uses a short, concise statement to explain its purpose: "COSI strives to provide an exciting and informative atmosphere for those of all ages to discover more about our environment, our accomplishments, our heritage, and ourselves. We hope to motivate a desire towards a better understanding of science, industry, health, and history through involvement in exhibits, demonstrations, and a variety of exhibits and experiences. COSI is for the enrichment of the individual and for a more rewarding life on our planet Earth."[7]

At the Science Museum of Minnesota in St. Paul, "Policy Guidelines" are utilized:

The newly expanded Science Museum of Minnesota is a strong commitment to excellence by the Museum Trustees. Through a generous dedication of funds and resources, we have established a goal of developing a nationally recognized institution for the purpose of interpreting the discoveries and insights of science for the general public and for identifying, collecting, protecting, and interpreting objects of scientific significance for future generations. In this endeavor, we believe the Science Museum will share the pride Minnesotans have in their other great cultural institutions and win a comparable following, probably even a much larger one because of the broad public interest in science and its social implications today.[8]

In Chicago, the Museum of Science and Industry relies upon a somewhat longer and more specific list of objectives for guidance:

1. To further public understanding of science, technology, industry, and medicine.
2. To explain scientific principles, technological applications, and social implications in an informal and enjoyable manner.
3. To preserve and interpret the nation's scientific, technological, and industrial heritage.
4. To acquaint young people with careers in science, engineering, industry, medicine, and related fields.
5. To present other appropriate cultural programs of general interest. . . .[9]

Amalendu Bose, retired director of the National Council of Science Museums in India, has stated that all science museums pursue the following objectives, with variations of emphasis: "to depict historically the stages through which science and technology has advanced in the country and also the world over; to portray the landmarks of scientific research and invention and to project the inventive genius of the scientists of the past; to depict through various museolographical means the fundamentals of science, its principles and phenomena, and the application of science for man's comfort and welfare; to provide a place for imparting informal education to young and adult alike in the field of science."[10]

In accomplishing these educational objectives, Bose has said, a science museum is required to arouse curiosity and stimulate interest in the minds of visitors about science, environment, and their interrelationship. "But are these objectives enough in the present-day context of the world?" he asks.[11]

Bose has pointed out that every social institution is a product of the environment and culture of its region and that museums are not an exception. Therefore, he asserts, the objectives of a science museum in a developing country will differ markedly from that of a developed country. It is necessary to appreciate the socioeconomic background of developing countries and the cultural moorings of their people:

Suffice it to say that the level of understanding about science and technology of the vast majority of the people in these countries is insignificant if not nonexistent. Obscurantism and superstition still persist in many regions; fruits of science did not percolate down to the bottom of the social ladder. To many, science is no more than an object of awe and wonder. Naturally, the people, by and large, are not receptive to new ideas or adaptive to new techniques.[12]

According to Bose, however, the state is committed in many developing countries to converting "this hapless lot of neglected people into a dynamic force for social change." This means, he says, that science, including science museums, must have relevance to the needs of society. "It has therefore become the obligation on the part of the science museum to undertake the responsibility of educating the masses—literate, semiliterate, or even illiterate—about the social benefits of science and the need to imbibe a value and practice a way of life imbued with scientific outlook."[13]

Such circumstances influence the differences in the objectives of science and technology centers in developed and developing nations. Another factor that affects institutional objectives is the nature and mission of the parent organization. The difference in two organizational patterns illustrates the point. In Baltimore, the Maryland Science Center is a division of the Maryland Academy of Sciences, which had its origin in 1797. The academy's objectives, as stated in its constitution, are to "promote the advancement of science; encourage public interest in and understanding of science; conduct an active program in science education, and serve as a central focus of scientific activities, both professional and amateur, in the State of Maryland."[14] The principal instrument for carrying out these objectives is the Maryland Science Center, a new science and technology center that opened in 1976.

In Japan, the activities of three science and technology centers— in Tokyo, Nagoya, and Osaka—are influenced by the purpose of the industrially oriented Japan Science Foundation: "to contribute to the improvement of the standard of science and technology in Japan by carrying out in an integrated and efficient manner various activities concerning the promotion of science and technology."[15]

In addition to helping to establish and support science museums, the foundation's activities include conducting surveys and making recommendations on science and technology; assisting organizations involved in science and technology; furthering cooperation between industry, universities, and public research institutions; disseminating information on science and technology; and operating a television station.[16]

An ordinance was passed by the City of Nagoya in 1962 to establish the Nagoya Municipal Science Museum. It listed the purposes of the museum as follows: exhibiting materials and apparatuses on science and technology; projecting the movement of celestial bodies by a planetarium and providing guidance in astronomical observations; operating a science library and presenting various research meetings, lectures, and motion pictures; conducting research on science and technology and publishing scientific materials; cooperating with other institutions to spread and promote scientific knowledge; and arranging other programs and undertakings recognized as necessary by the mayor.[17]

Implementing the Objectives

Objectives are of little value unless they are implemented. Too frequently, a list of objectives adopted at the founding is forgotten. As a result, the institution suffers from the lack of purpose or direction.

Objectives should be adopted and applied and even modified when outdated. They provide the foundation for a science center's activities and community interest. They keep an institution from straying from its designated purpose.

A clearly defined role also can be extremely helpful in planning the future of a science center and in obtaining the necessary financial support for its operation.

6

The Governing Board

Virtually all science and technology centers have a governing board. About the only exceptions are institutions that are not independent corporations but are controlled by the governing body of a parent organization or by a committee, commission, or council appointed by that body.

A science center's governing board normally is called a board of trustees because it is entrusted with the responsibility of operating the institution on behalf of the community served. Sometimes this body is known as a board of directors.

A board of trustees has two basic functions: to establish the direction and basic policy for the management of the science center and to adopt a financial program that will provide sufficient funds for the operation of the institution. Among the policy matters usually decided upon by a governing board are the overall objectives, organizational pattern, types of programs, collection policies, building expansion, investments, and long-range plans. A board also hires the director, approves the budget, raises funds, and sometimes makes decisions concerning exhibits, education programs, use of facilities, and key personnel.

The financial role of trustees has become increasingly important. In approving the budget and expansion plans, the governing board commits itself to raising the necessary funds to fill the gap between income and expenses. Few science centers have balanced budgets without the aid of contributions made or solicited by trustees.

The specific functions of a board of trustees generally are spelled out in an institution's constitution and/or bylaws. However, these basic legal responsibilities frequently are augmented by other duties resulting from a science center's needs, personalities, and/or traditions.

In an essay on museum trusteeship, Helmuth J. Naumer, executive director of the San Antonio Museum Association, wrote:

A museum's working foundation is an understandable, flexible, and up-to-date set of rules, *i.e.,* a constitution, by-laws, and rules for day-to-day operating procedures. Regrettably, many museums operate under hopelessly outdated guidelines. Many policies were conceived before the Code of Hammurabi—at least they appear that hoary—or when a board was new and naive or when survival was the only concern or when an individual was trying to feather his own nest. Guidelines should be established by the entire board of trustees after due deliberation and ample opportunity for all viewpoints to be heard.

These guidelines need not serve as a straight jacket, but they should not contain the sort of legalistic jargon commonly seen in charters. The museum's field of interest should be stated in terms clear enough to be used as a yardstick for the consideration of proposed activities. The constitution and by-laws, whether long established or recently written, should be reviewed to be sure that they incorporate the true purpose of the institution, that they reflect current conditions, and that they establish a frame of reference in which the various entities of the museum have their proper place.[1]

Selection, Organization, and Operation

The selection, organization, and operation of a governing board usually are covered in a science center's constitution and/or bylaws.

There is no optimum size for a board of trustees. What works in one community will not necessarily apply under different circumstances elsewhere. Governing boards range from 6 to nearly 100 members, with 20 to 40 being most common. A small board can be insufficient to deal with the range of activities at a science center, while a large board can be unwieldly and have too many inactive members.

Some boards also have ex officio and honorary members. For example, the Museum of Science in Boston, in addition to its 76 members, has 7 ex officio and 6 appointed members. The ex officio members of governing boards usually are city, county, and/or school officials. In Boston, they also include the governor and two senators. Honorary trustees normally are retired board members or persons cited for their contributions to the institution.

Some science and technology centers, such as the Pacific Science Center in Seattle and Ontario Science Centre in Toronto, also have an advisory council or citizens' board, which makes it possible for them to solicit the participation of a large number of leading figures without imposing the responsibilities of the board of trustees on them. Such a group sometimes consists of 50 to 100 members. The principal purpose of such an advisory group is to strengthen community interest in and support of the institution.

In general, board members are elected by the voting members of the nonprofit corporation that operates the science center. The board may comprise the only voting membership, or voting privilege may be based on the level of dues a member pays. Some board members may be appointed because they hold other positions, such as the presidency of a women's or volunteers' organization.

The term of office most frequently is three years, although two-year terms also are common. The terms usually are staggered so that one-third or one-half of the board is retired each year. Board members may or may not serve successive terms. Many institutions also have a retirement age—generally sixty-five—for active board membership as a means of fostering new leadership.

The composition of a board of trustees will vary greatly. Some science centers have trustees that are largely civic, corporate, educational, and scientific leaders, while others seek to have as many community interests as possible represented on the board. Some institutions have a board consisting entirely of volunteer workers. Such an approach, however, limits the potential of the science center. It needs the leadership and expertise of the business executives, professional people, merchants, scientists, educators, media specialists, civic and social leaders, and others, providing they are interested in and willing to work for the science center. An increasing number of science centers also are including women and minority representatives on the board.

An ideal board member, according to one trustee handbook, is one that has "strength, vision, experience, and a blending of educational, managerial, investment, business, and legal talent."[2] A few more desirable attributes would be dedication, willingness to work, and ability to raise funds. Unfortunately, few trustees have all these qualifications.

Too often, people are selected to serve on the governing board purely as an honor or because of his or her position, without any obligation to contribute to the betterment of the institution. It is vital that board members have a sincere interest in the museum and a willingness to accept the responsibilities that go with the position. This includes helping with the financial support.

A board of trustees normally meets a fixed number of times per year, usually every two or three months. Board members should be required by the bylaws to attend a minimum number of meetings or be subject to removal.

The key to a successful governing board is a competent and conscientious chairman. He or she must be a leader in the community who is willing to devote the time and effort necessary to guide and support the institution without becoming a problem for the director and staff.

Officers usually are elected annually. In addition to the board chairman, a science center may have a president, one or more vice presidents, a secretary, and a treasurer, and nearly always has an elected or appointed executive committee to act on behalf of the board of trustees between meetings.

Most governing boards have standing or working committees to assist in discharging its responsibilities. These committees usually deal with such areas as finance, collections, promotion, building, audit, and nominations. They normally are appointed by the president and are required to report periodically on their activities. The finance committee generally is chaired by the treasurer and has the essential job of watching expenditures and the budget, managing investments, and obtaining new sources of support when needed.

Helmuth J. Naumer, in his booklet on museum trusteeship, suggested that both new and incumbent trustees be provided with an annually revised handbook containing such information as museum general information and statistics, board and staff organization chart, museum constitution and bylaws, trustee list with addresses, policies and amendments of the board, personnel regulations, etc., resumes of professional staff, staff roster with titles and addresses, agreement with trustees and director, and current annual report and budget.[3] Although few museums have such a handbook, Naumer believes it would be helpful in orienting board members and improving their understanding of the institution's objectives, operations, and needs.

Potential Board Problems

A governing board can function smoothly and contribute greatly to the success of a science and technology center. It also can be a problem and have problems.

A delicate balance must be maintained between providing assistance and meddling in the operations of the institution. As pointed out at a recent conference, "Board involvement in the daily operation of the institution is a difficult problem. . . . Most professional staffs feel that this is an invasion of their professional prerogatives and that the staffs should be left free to operate within approved policy and financial guidelines. However, board awareness of staff projects and general museum programs is essential."[4]

A reverse problem is when the director and staff usurp the authority of the board of trustees. Basic policies concerning the direction, use, and support of the institution are clearly the responsibility of the trustees. The staff must be careful in initiating new programs that stay within board policies. It also should be remembered that the director is a salaried employee of the board of trustees charged with the administration of the affairs of the museum in accordance with the policies and within the budget limitations established by the board.

The size, nature, or composition of the governing board may prevent the board from having a quorum for meetings or getting action on important matters. The use of an executive committee, a standing commiteee, or an ad hoc committee frequently can overcome such a problem.

Sometimes a governing board is hampered by a trustee who becomes an obstacle to orderly progress or becomes inactive. The board should have a built-in mechanism to deal with such problems, such as a fixed term of office, a prohibition on multiple terms, a retirement age, and/or an honorary trustee category.

It frequently is helpful to have a banker, a lawyer, a businessman, and other such knowledgeable people on the board of trustees. However, a science center and its board members always must be on guard against conflicts of interests. Although most trustees are diligent and honest, extreme care should be taken to avoid the slightest hint of inpropriety in depositing funds in the bank of a board member, utilizing the legal services of a board member, or purchasing supplies from a board member. Board members should abstain from participating in decision-making discussions when they involve their services. Whenever possible, the decision should be based on closed bids, or the trustee's firm should be eliminated from consideration.

The acquisition and disposal of collections and fiduciary responsibilities also are areas of potential conflict of interests. A science center must exert extreme care in buying, accepting, selling, trading, and/or giving objects of intrinsic value when trustees are involved. "The fiduciary responsibility of boards is becoming more critical and subject to review," according to the Spring Hill Center Museum Conference. "The management of portfolios of endowments and pension funds is coming under closer government scrutiny and regulation."[5]

The investments of endowed museums typically have been managed by a financial institution represented on the board of trustees. As regulations for performance review of money management and accountability become stricter, the best route may be to avoid any question of conflict of interests by not having the investment firm sitting on the board.

John Henry Merryman, professor of law at Stanford University, has pointed out that trustees have "an obligation to exercise reasonable care and diligence in their deliberations and actions. Failure to meet this 'duty of care' is negligence and exposes the trustees to legal liability. It also constitutes a violation of their social obligations."[6] He said the courts have been generous to trustees in the past by generally characterizing their actions or inactions as falling within the permissible range of care. "However, there are limits, and if negligence is found, the full range of legal remedies against the trustee potentially is available, including personal financial liability of the trustee, removal of the trustee from his position, and, in extreme cases, criminal liability."[7]

It is for such reasons that many science centers have taken steps to avoid any possibility of conflict of interests and have purchased fiduciary insurance for the board of trustees.

7

Organizational Structure and Staffing

Organizational Framework

The organizational structure of a science and technology center will vary with its degree of autonomy, its institutional size, the nature of its activities, and the preferences of the governing board, parent body, and/or director. However, despite these factors, there is considerable similarity in organizational patterns.

It usually makes a difference whether a science center is an independent nonprofit institution, a part of another organization, or an instrument of government. This degree of independence frequently determines whether it has an elected governing board with far-reaching policy and management responsibilities or an appointed committee, commission, or council with limited powers.

The size of a science center also affects its organizational structure. A small institution, for example, may require only a director and a few supporting personnel, while a large science center may need many more staff members and a much more elaborate organizational framework.

The organizational structure frequently is influenced by the nature of an institution's activities. A contemporary and participatory science center often requires different types of people and a different organizational pattern than a collection-oriented institution.

Another factor in the organizational picture is the human element—the people who make up the governing board or parent body and the person who is the chief administrator of a science center.

Regardless of other influences, it is possible to have an entirely different organizational mix based on personal preferences.

In general, the organizational framework of science and technology centers can be grouped into three broad categories based on the size and nature of the institutions. Small participatory science centers generally have a governing board, director, secretary-bookkeeper, exhibit designer-builder, education-public relations director, guide-guard, janitor-handyman, and volunteers to assist with visitors and operate the store.

The operation of medium and large participatory science centers involves a governing board; director; assistant director; departments concerned with business affairs, exhibits, education, and public relations, and offices or departments dealing with security, building maintenance, membership, fund raising, library, and volunteers.

Collection-oriented science centers may be similar to either of the foregoing categories, with the addition of one or more curators of collections, a clerk-registrar, and possibly one or more specialists for restoring and preserving objects in the collections.

The Director

The principal administrative officer of a science and technology center usually is called the director. In small institutions he or she sometimes may be called a curator; in large science centers, the title of executive director, director general, or president may be used.

The director plays a pivotal role in the success or failure of a science center. It is the director's responsibility to administer the policies of the governing board, to manage the day-to-day operations of the institution, and to provide the leadership necessary for community interest and support.

The director is a paid, full-time employee selected by the governing board. He or she usually hires and fires the staff, develops and abides by the budget, and is responsible for all aspects of the institution's operations, including collections, exhibits, programs, planning, and fund raising.

A director must be able to manage people, programs, and funds; to work with the schools, members, contributors, and trustees; to articulate the institution's objectives and needs; to develop an interested and supportive constituency; and to demonstrate imaginative thinking in dealing with the institution's needs.

Museums have been compared to a business. "Nowhere is the similarity more apparent than in the area of museum administration. Accepted business practices should be instituted to govern financial matters, operations, and employee relations," according to one museum planning guide.[1]

In a small science center, the director must be a jack-of-all-trades. The director may be the only professional on the staff. In a larger institution, he or she must be able to direct the efforts of various specialists for maximum effectiveness, possibly with the aid of an assistant or deputy director. Either way, the position requires a highly competent and resourceful person who is becoming increasingly difficult to find in the job market.

At one time, the director of a science museum generally was a former curator who was an authority in some field of science or technology. With the development of contemporary and participatory science and technology centers, it is no longer necessary for a director to be a subject-matter specialist. A director also may be a professional in education, communications, or business management. The field, however, is not as important as the training, experience, and talents brought to the position. Whenever possible, a director should be hired who has had experience in the museum field, preferably as a top administrator or department head at a science center.

Unfortunately, there are no schools that give adequate training in museum management. Therefore, in starting a science center, it usually is necessary to entice a staff member away from another institution or find a person in the community with some of the requirements and a missionary zeal who can receive on-the-job training with the guidance of the board of trustees.[2]

In his guide to museum management, Carl E. Guthe said a museum director is "the personification of the museum."[3] Putting it another way, a governing board must diligently search for the best applicants and make the right choice because the director *is* the museum to many people in the community.

The Staff

The number and types of staff members at a science center will fluctuate with the needs of the institution. However, regardless of the size or nature of the staff, most of the basic functions described below must be performed.

Business Department

In a small science center, the financial function may be handled through the director's office with the aid of a secretary and/or bookkeeper. More often, however, there is a business or administrative department with several staff members.

The chief fiscal officer is the business manager. He or she is usually an accountant or someone with experience in business administration. In larger institutions, there also may be other accountants, a payroll clerk, and even a personnel manager.

The business department normally is responsible for fiscal transactions, budgeting, payroll, personnel records, insurance, retirement benefits, sales operations, and financial reports. In some cases, the operations, security, building maintenance, development, and/or membership functions also are part of the business department. When this occurs, a person generally is assigned to supervise each function.

Curatorial Department

Most science centers have collections of objects of intrinsic value, but relatively few have curators and supporting personnel to gather, restore, preserve, study, and interpret the collections. At most science centers, the collection, record-keeping, and maintenance functions are performed in the business, exhibits, and/or education departments. Some institutions use outside personnel on a part-time voluntary or contract basis for restoration or conservation.

A curatorial department, when it exists, normally has one or more professional curators who may be scientists, engineers, historians, or people knowledgeable in the subject matter and/or the care of the collections. Sometimes curators are assisted by technically trained personnel skilled in the preservation of objects.

Exhibits Department

Because of the importance of exhibits at science and technology centers, the exhibits department usually is one of the largest and most active of the departments. In most cases, the head of the exhibits program is known as the exhibits director, but in small institutions, he or she might be called the exhibits designer or coordinator.

The exhibits department is charged with developing and maintaining permanent exhibits and scheduling and handling temporary and traveling exhibitions. This may involve designing and constructing exhibits internally and/or contracting with outside designers and builders to do the work. Frequently the exhibits staff also is responsible for museum graphics, publications design, and caring for the collections.

In a small institution, the exhibits department may consist of one or two people. A large science center may need fifty or more on the staff. An exhibits director and key department personnel usually come from design or architectural backgrounds. Other departmental personnel will vary from center to center but may include draftsmen, electronic technicians, projectionists, electricians, carpenters, machinists, and painters.

Education Department

The education department is responsible for coordinating all of the institution's educational activities, from handling school group visitations to preparing interpretive materials. The education director, or supervisor, frequently is a former science teacher.

Among the many concerns of the education department are exhibit tours, science demonstrations, study materials, teacher handbooks, school newsletters, educational courses, lectures, films, field trips, teacher institutes, nature trails, science kits, outreach programs, summer camps, and other educational activities.

In a small science center, the education director may be assisted by a number of volunteers or docents. In a large institution with many courses and other programs, it may be necessary to supplement a staff of five or more with outside lecturers, teachers, and field trip leaders.

Public Relations Department

Promotion is a necessity for any science center. It is important to keep the trustees, employees, members, volunteers, contributors, exhibitors, city officials, editors, and the general public informed of the institution's activities, needs, achievements, and aspirations. This must be done with or without a public relations department, but it generally can be accomplished better with a trained communications staff.

The public relations director—sometimes known as the public information director or public affairs director—usually has a media background, with newspaper, magazine, radio, and/or television experience.

The public relations department's activities normally include preparing news releases, newsletters, guide booklets, direct mail, radio and television announcements, posters, and promotional films and coordinating television appearances and a speakers' bureau.

Other Departments and Offices

A number of science center functions are handled in two ways, either as separate departments or as offices within other departments. They include operations, security, building maintenance, development, membership, and volunteers.

Operations

The day-to-day operations of a science center, such as handling the visiting public, security, and programs, sometimes are administered through an operations or programs department, headed by an operations, program, or administrative director. More often, such functions are the responsibility of the director or the business manager.

Building Maintenance

A building engineer or building manager normally has charge of the building maintenance. He or she may be a janitor-handyman in a small science center or a professional engineer in a large institution. The maintenance crew may include janitors, electricians, carpenters, and heating and cooling specialists.

Development

The development office is responsible for raising funds, but at many institutions the function is carried out by the director or business manager. Normally, the office keeps the fund-raising records, directs solicitations, and handles events and people with contribution possibilities. A development director or officer may be an experienced fund raiser or a socially active individual with organizational skills and contacts.

Membership

Most science centers have a membership program, but not all have a separate membership office. A typical membership office is concerned with seeking and serving members as well as keeping the membership records. This involves preparing membership literature, sending out mailings, collecting dues, organizing membership programs, and sometimes conducting annual giving drives. A membership director also is known as a membership secretary or coordinator. The position usually calls for a personable individual with good organizational and communications skills.

Volunteers

Volunteers are handled in many different ways. Sometimes they are administered through a separate office, while other times they are part of the education department, membership office, or women's board. A volunteers coordinator usually has administrative responsibility for organizing, training, scheduling, and monitoring volunteers. This person may be a volunteer or paid employee with organizational and supervisory abilities. Volunteers who conduct guided tours and give lectures are known as docents.

Adjunct Groups

In addition to membership and volunteer organizations, some science and technology centers have support, service, and/or advisory groups affiliated with their institutions. The three most common approaches are the friends' organization, women's board, and citizens' advisory council.

Friends' Organization

Friends' organizations are primarily support groups. Like member organizations, they consist of interested people from the community who provide funds, materials, and personnel for the ongoing operation of the institution and special projects. Friends' organizations frequently are found at science centers that are not autonomous and find it difficult to have membership programs because they are operated by governments, school systems, universities, or other parent bodies.

Women's Board

Before the women's liberation movement, it was common to have a separate women's board or auxiliary at museums. Although less frequent now, some institutions still have such a mechanism for women to take an active role in the organization. Some women's groups focus on the support of a single project, and others provide volunteer personnel and funds on a continuing basis for the operation of the science center.

Citizens' Advisory Council

Sometimes it is desirable to involve a large number of leading citizens in the affairs of a science center without charging them with the responsibilities of a board of trustees. They may form a citizens' advisory council or board that meets several times a year to hear about the institution's activities, needs, and/or plans. Such a group may consist of fifty to one hundred or more people who are approached for suggestions and assistance from time to time. Some science centers also make use of other forms of advisory groups for planning exhibits and programs.

8

Management and Planning

Elements of Management

The day-to-day management of a science and technology center frequently is overlooked in the concern for collections, exhibits, and programs. But without adequate direction, organization, and/or planning, a science center can easily encounter financial, programming, and operation problems with serious implications for the institution.

The overall management of an institution is the responsibility of the museum director, but the administration of a science center also involves the guidance and support of the governing board and the input, cooperation, and follow-through of department heads, supervisors, and other staff members. Running a science center is a team effort, with the director serving as the coach. He or she must provide the leadership, decide on the operating program, and see that the game plan is implemented.

"Management" is one of the least understood and most poorly practiced functions at museums. The term frequently is used to refer to the people at the top and/or to those people who direct the work of others, but management is a process. In an article on the management of cultural organizations, Frederick J. Turk stated that the trustees and administrators of museums and other arts groups must use the management process "to utilize scarce resources most advantageously." He said the process typically consists of planning, implementing plans, monitoring progress, and evaluating results.[1]

Peter F. Drucker, the noted management consultant and writer, has said that management has three basic jobs: "managing a business, managing managers, and managing workers and work." He also pointed out there is an additional major factor "in every management problem, every decision, every action" that becomes a fourth dimension of management: time.[2] These management functions are closely related and cannot be separated. "Any management decision always affects all three jobs and must take all three into account," advises Drucker. "And the most vital decisions on the future are often made as decisions on the present."[3]

Managing a museum is somewhat different from managing a business, but many of the same factors—goals, organization, people, programs, funds, evaluation—are involved. The principal objective may not be to make a profit, but one of the management goals certainly must be to avoid a deficit. Basically, it is the responsibility of management to operate an institution in the most effective and efficient manner consistent with the institutional objectives.

A "manager" has two specific tasks that nobody else discharges, Drucker stated. "The manager has the task of creating a true whole that is larger than the sum of its parts, a productive entity that turns out more than the sum of the resources put into it. . . . The second specific task of the manager is to harmonize in every decision and action the requirements of immediate and long-range future."[4]

The five basic activities of a manager, Drucker explained, are to set objectives; organize; motivate and communicate; measure performance; and develop people.[5] These managerial responsibilities are just as meaningful in a museum as in a factory and at the bottom as well as the top of an organizational structure.

The management of a science and technology center involves many elements, some of which follow.

Objectives and Goals

Every science center should have broad institutional objectives, and every management team should have specific goals. Management needs a plan and goals against which to measure its performance. It is the museum director's responsibility to provide the goal-oriented leadership on the basis of the objectives, needs, and resources.

Board Support

To be effective, the management of a museum also requires the support of the board of trustees. In delegating the administrative responsibility to the director, the trustees should not meddle in operating affairs. The board should back up management in its efforts. However, when the trustees no longer have confidence in the director, they should make a replacement.

Administrative Structure

Management is not a one-man band. It is more like an orchestra with the museum director as the conductor. Thus, a science center needs a suitable organizational structure to plan, organize, and implement its activities. The nature of this administrative framework should be appropriate for the institution's purpose and size and the director's mode of operations.

Staff

For any management plan to work properly, a director must have competent people on the supporting staff. As indicated earlier, the management staff is involved in setting goals, organizing, motivating, communicating, measuring, and developing people. This requires capable people in all management areas, regardless of the structure.

Programs

The main reason for the existence of any museum is the programs and services it provides, and the principal role of management should be to see that these activities are furnished in the best possible way within the constraints set by the institution. The objective of management should not be to perpetuate management.

Funds

A science center needs funds to pay the bills. It also is necessary for the director and other management people to assist in raising the funds and to make judicious use of available resources. Money is not the whole answer to an institution's needs, but it is essential to its financial stability and programming. It also costs money to raise money, and the governing board and management should recognize the realities of fund raising.

Employees

The needs and supervision of employees must be considered in any management plan. Are the lay levels and fringe benefits adequate? Should the museum have an orientation program, supervisory training, job descriptions, pay grades, safety regulations and training, rules of conduct, a grievance procedure, and/or an administrative or general orders manual? Management also must be concerned with the impact of such activities on unionization. Relatively few museums have unions, and most of them are located at institutions operated by municipalities.

Planning

A vital tool of management is current and long-range planning. Budgeting is part of this process, but it should come after the goals and programs are settled. Too frequently, the reverse is true. Planning should be related to a science center's objectives and needs, and an attempt should be made to raise the necessary funds for programs and services that are a result of solid institutional planning.

Evaluation

Monitoring and evaluating programs and other activities should be part of any management and planning process. It is essential to check on performance and to evaluate the results. Monitoring involves periodic comparison of the implementation effort and the plan, while evaluation determines whether the effort was worth it. Both are extremely useful in planning future activities.

Management Control Systems

In most fields, management control systems play an important role in identifying, reporting, and evaluating organizational performance. This has not been the case with museums. In general, museums have emphasized professional quality rather than cost control. Internal and external reporting systems have been largely unsophisticated. But with increasing financial demands, it has become necessary to make optimum use of existing resources and to generate new revenue.

The key to good management is an effective set of management control tools. It is for this reason that the National Endowment for the Arts contracted with Management Analysis Center Inc. (MAC) to study the application of such management techniques in museums. Approximately one hundred museum professionals were interviewed, and a museum advisory committee was utilized in the process. The result was a helpful report, *Museum Management Tools,* published in 1978.[6]

The study sought to develop a system of management tools tailored to the needs of museum managers. The tools basically evaluate performance by setting standards and obtaining relevant and timely information for decision-making.

The report stated that a system of management tools "should be designed to support the museum's mission and help the museum achieve the goals and performance standards it set for itself."[7] These goals and standards are set, according to the MAC study, through four management decision-making processes, which it called among the most important of management's tools: strategic planning to determine long-term options and goals; tactical planning, which translates strategic plans into more concrete, medium-term plans and performance targets; operational planning and budgeting, which establish one-year activity implementation schedules in support of plans and utilize current resources within that context; and financial planning, which seeks to obtain the necessary capital to carry out plans and strategy.[8]

It is through such planning processes that management identifies factors that may have a significant influence on how well the museum is able to reach the goals set by management. Management then selects from this group those factors it believes are essential to the museum's success, such as attendance, fund raising, or elements of the exhibit, membership, or education programs. These factors provide the framework for goal achievements at the museum.

In those areas where management is able to control key success factors, it should set standards of performance and communicate the goals to those people responsible for meeting them; measure performance against standards; provide decision-makers with relevant and timely information on progress; and present options for corrective actions when the standards are not met.

In those important areas that management cannot control, the report recommended that management identify programs, activities, and services subject to significant uncontrollable influences, such as weather, tourism, and inflation; provide information about these factors and their potential effects on museum goals and activities; and propose alternative courses of action in case the uncontrollable factors adversely affect operations.

For both controllable and uncontrollable factors, timely and meaningful information is a critical element in the management process. MAC listed five types of management reports that are helpful in the process: overall museum strategic and tactical performance against goals, operational performance against plan and budget, financial performance against plan and budget, allocation of critical resources by strategic classifications, and trends in external factors important to performance and goal achievements.

Strategic, tactical, and operational planning are a necessary part of the management control process. Strategic planning sets the general direction of the museum. By stating the museum's mission, the board of trustees and/or museum director identifies a target image, audience, and community and provides guidance in the development of museum strategy. In developing strategy, management determines the range of possible future activities and programs available. The effect of each strategic option on a museum's present and future audiences and its professional and financial standing then are evaluated, and the most viable option is selected.

Tactical planning translates strategy into specific goals and standards for performance over a two- or three-year period. Unfortunately, medium-term planning frequently is done on a program-by-program basis rather than on a museumwide basis within the framework of a strategic plan. As a result, single-program planning often fails to provide overall direction, coordination, or sufficient consideration of how to obtain the necessary resources.

Operational planning and budgeting expresses and quantifies programs in terms of current year's activities. The budget summarizes goals, lists programs to be undertaken, and set targets for admissions, sales, programs, services, costs, revenues, and cash position. The budget forces management and staff to define planned programs and services realistically in terms of available resources and communicates these expectations to key personnel through quantified standards of performance for the year.

Museum management has a variety of tools that can be used to measure how well the operations of the museum are achieving the predetermined standards. The NEA-sponsored study pointed out that the performance measurement tools should:

1. Provide a capsule report on actual vs. predicted performance.
2. Focus on the key focus factors identified in the planning process.
3. Indicate options for management action if variance occurs in controllable success factors.
4. Provide information usable to modify strategy, plans, and standards if variance occurs in areas beyond the control of management.[9]

In conducting the interviews with museum representatives, MAC identified four major categories of activity that are essential to museum success: overall mission and strategy, revenue, development, monitoring of costs, and ancillary services.

MAC concluded its report by saying that

In the end, good management is sound decision-making based on good information. The tools presented in this report facilitate decision-making by providing sound information. They are based on management practices observed in museums across the U.S.— many with systems far more complex than those presented here, and others with more simple systems. The majority of performance measures make specific a director's intuitive judgment, thus easing the management role by permitting data aggregation at a lower organizational level and by providing documentation for reports to the Board of Trustees and external groups. Regardless of the system selected, the judgment and common sense of the director and his management are the determining factors in the success of any system.[10]

The Planning Process

As demonstrated in the foregoing section, planning is an integral part of the management function. It should be a continuous process that covers both the immediate and distant future. Good planning usually produces both good budgeting and programming.

"To ignore planning," a Canadian Museums Association management guide cautioned, "is to proceed by rote or by guesswork, and it is hard for a professional to justify guesswork. Having a plan before plunging into work is like furnishing oneself with a road map before setting out to travel an unfamiliar highway."[11]

To effectively marshal the resources of a museum and a community, the director must work toward specific goals. Short-term goals usually are developed by the museum director in collaboration with the management staff and then approved by the governing board. For long-term goals covering three, five, or ten years, the trustees should be involved from the beginning because of policy and financial implications.

Most museums do not have professional planners on the staff. The director tends to use the business manager and other department heads for the preliminary research and discussion. In some cases, an architect or planning firm may be hired to supplement the internal planning, especially if it involves new facilities or exhibits.

H. Lawrence Wilsey, a vice president for the Booz, Allen & Hamilton Inc. management consulting firm, believes planning is needed to stave off "financial disaster and organizational collapse" in many organizations, particularly new ones.[12] A long-range plan, he stated, calls for basic decisions at each of seven planning levels: institutional philosophy, objectives, programs, organization, staffing, facilities, and finance.[13] The following are some of the typical questions Wilsey says should be asked.

Philosophy

What should be the role of the organization in meeting the cultural needs of society and the individual? What knowledge, skills, and interests need to be developed to meet these needs?

Objectives

What are the expectations of the people of the community? To what extent should the organization strive to meet or modify them? To what publics should the museum appeal? Are the cultural needs of these publics being satisfied by other organizations? If so, how should the institutional objectives be changed? What levels of audience exposure should be sought in the future? What should be the membership goals?

Programs

What should be the type and range of programs to achieve the institution's objectives? What program effort should be directed toward school-age children? What effort should be aimed toward groups with extensive leisure time?

Organization

What abilities are required to execute the organization's programs? What type of board, staff, and committees are needed to plan, conduct, and evaluate the work of the organization? What should be the assignment of functions and responsibilities? What working relationships and reporting responsibilities should make the museum's work most effective? How should external relationships be conducted?

Staffing

How many people with what qualifications and characteristics are needed to provide effective board, professional staff, and volunteer leadership for the organization? What kinds of incentives can be given to volunteer workers? What type of classification, salary, and professional development program is required?

Facilities

What number, kind, quality, and location of facilities are required to serve the defined program needs? Are the proper types of facilities being planned for future expansion, programs, flexibility, and economy?

Finance

What operating and capital funds will be required to support the programs and provide the necessary staff, facilities, equipment, and supplies? When will these funds be required? From what sources can they be obtained? What financial programs can be modified? Which can be extended if necessary? What type of fund raising is appropriate for the organization?

Planning is a demanding process. It requires considerable effort, time, and information—and sometimes funds. But there are benefits, as listed by Wilsey: better definition of the organization's objectives, with assurance that its goals are in reasonable harmony with the community's expectations; more effective programs; a higher level of morale and enthusiasm by those who feel a part of the organization and its future; a sounder approach to financing programs, with financial goals keyed to achievable and desirable program levels; and greater assurance that the organization is contributing effectively toward meeting cultural needs of the community.[14]

Malvern J. Gross Jr., a specialist in nonprofit accounting at Price Waterhouse & Company, believes one of the most effective ways for any nonprofit to avoid the unexpected is to prepare and periodically update a five-year master plan. "The purpose of this five-year plan is to force the board to look ahead and anticipate not only problems but goals and objectives that they want to work toward achieving."[15]

He pointed out that management can initiate and push the board toward developing a long-range plan, but the board should be involved because any plan of action must be developed by all the people who will have to live with the resulting plans.

It is possible to approach the planning process in a number of ways. One method might be for the board chairman and/or museum director to appoint a committee of three or five people who have key policy-making or management roles. Gross outlined the suggested planning procedures as follows:

Setting Goals

Each member of the planning committee should take five sheets of paper—one for each of the five years. On each sheet, he or she should write the goals that are seen as being important for that year without regard for cost. The goals can be general or quite specific. Then the committee should discuss the various projections until they can agree on a reasonable list of five-year goals.

Estimating Costs

Once the committee settles on the goals, it is appropriate to estimate the costs involved in reaching each goal. This can be difficult because of the many unknowns, such as inflation and costs at the time of implementation. The treasurer or business manager should be given the task of obtaining estimates with the assistance of other staff members and outside specialists. After the estimates have been made, they should be totaled.

Plan for Income

The final step in the long-range planning process is to determine how the necessary funds will be raised. This assumes, of course, that the board of trustees approves the plan. It may decide to modify portions of the plan because of costs or disagreement over the goals.

"The importance of planning into the future cannot be over-em-phasized," according to Gross. "In this fast-moving age, worthy nonprofit organizations can quickly get out of step with the times, and when this happens, contributions and income quickly disap-pear. A five-year plan is one technique to help ensure this won't happen."[16]

9

Financing and Fund Raising

The Case for Support

Regardless of the monetary sources or the fund-raising techniques, a science and technology center must present a convincing case for financial support. Its mere existence is insufficient justification for its survival.

Whether it is a new or old institution, a museum has a selling job. To receive the funds it needs to operate, it must convince the public, government officials, wealthy individuals, companies, foundations, and others that it merits support in the form of contributions, admissions, tax funds, membership dues, and research grants.

Too often new science centers are launched without an adequate financial base or sufficient fiscal planning. Existing institutions sometimes experience financial difficulties because of poor fiscal management and an inability to raise enough funds to keep pace with rising costs. This chapter deals with developing sources of support; the next one is concerned with financial management. Both funding and management, however, depend upon the case for support.

In some ways, the arguments for support are similar to an institution's purpose and objectives. But they go beyond goals to point out specific services or contributions made by the science center to the community, region, or nation. This case for support must convince the constituents that the cost is worth the return, whether it is a publicly or privately funded museum.

Increasingly, museums of all types are depending on a variety of sources for support. In seeking funds in a highly competitive environment, they must demonstrate their value and justify their funding. There are too many worthwhile appeals for support to assume that funding will be forthcoming automatically or logically.

At many museums the long-standing reliance on governing board members to raise necessary funds is changing. Financial needs are increasing and traditional funding sources are decreasing. Professional fund raisers are being hired, and planned development and public relations programs are being instituted.

"An adjunct of the move toward professional fund-raisers is the need for museums to make their case more clearly not only to the public, but to municipalities, counties, and states," according to the Spring Hill Center museum conference. "A professional public relations program will be essential in planning and carrying out the information needs in addressing all sources of support, particularly local and state governments."[1]

Like all museums, science and technology centers need more diversified and reliable funding sources. As one museum planning guide has pointed out, "A museum which depends entirely for existence on an annual membership drive or on special events is living 'hand to mouth.' For lasting financial security, institutions need more dependable sources of income for basic needs."[2]

A balance of support from existing and new sources requires a game plan and a convincing case for annual, capital, and deferred giving from trustees, members, and friends; contributions from industry; grants from foundations; large attendance and admissions from the public; and increasing support from the tax coffers.

The arguments for support generally deal with an institution's role in the area served. This information should be transmitted through media, publications, speeches, audiovisual presentations, meetings, and conversations. A typical institutional case for support might include such points as how the center helps to further public understanding of science, technology, industry, and health; what role it plays in preserving and interpreting the region or nation's scientific and technological heritage; how it fulfills a vital need for supplemental science education for children and adults; how it serves to inspire and assist youngsters interested in careers in science, engineering, and medicine; how it increases the economic security of the community it serves; how it contributes to social uplifting through its programs for minorities, senior citizens, the handicapped, ethnic groups, and other segments of the community; and how it informs citizens about scientific and technological issues of great importance to the community and society in general.

In some cities, museums and other cultural organizations have sought to document the economic value of their institutions. A joint study by seven museums revealed an economic impact of $228 million annually in Chicago and a similar amount in the suburbs through purchases by visitors from outside the metropolitan area.[3] In Philadelphia, a survey found that an additional one hundred visitors daily to the city's cultural organizations generated $78,000 in taxes, $1.2 million in sales, and the equivalent of 111 new industry-related jobs.[4] A Minnesota study disclosed that 233 responding nonprofit organizations poured $20 million each year into the state's economy, with about half going for payroll and the remainder for purchased supplies, equipment, and services.[5]

Whatever points are emphasized, a compelling case for support should be made. The development of such a statement also can be useful to founders, trustees, and administrators in planning a new institution, improving an existing one, and in determining needs and making long-range plans.

Sources of Support

Despite the differences in science and technology centers, all have the same basic need: a dependable source of support. Without adequate funds, it is difficult for any museum to function effectively, regardless of the staff, programs, or intentions.

To the public, it may seem as though museums are poverty-stricken most of the time. As G. Ellis Burcaw pointed out, "This is probably because the public expects more—and their professional staffs desire to give more to the public—than their operating budgets will buy."[6]

It is an unusual museum that can support itself without outside financial help. Being nonprofit institutions, their sources of earned income are limited, and it is necessary to rely on public and private support to produce a balanced budget.

Unfortunately, it is becoming increasingly difficult for all types of museums, including science centers, to operate without deficits.

They are being called upon to do more, frequently with less, considering inflation, staff, and other factors.

In many countries, museums are operated and funded by national, regional, or local governments. The failure of government authorities to increase appropriations sufficiently to come with mounting costs is a growing problem in such cases. In the United States and those countries where museums are largely privately controlled and financed, traditional sources of support are being eroded. Thus, it is necessary to develop new and expanded sources of funding for institutions in both systems.

Funds for science centers and other museums take two forms: revenues and support. Revenues are earned income generated internally, while support comes from external public and private sources. The funding of most institutions is a mix of both.

Nearly every science center needs a greater financial base and is working to increase earned income and outside support. But the nature of the institution, community, and/or government may restrict the possibilities for certain types of funding. Therefore, in planning a new science center or expanding the support of an existing one, a careful and realistic examination of the financial potential is a necessity.

Revenues

A science and technology center can earn income in many ways: investment income, admission fees, membership dues, program fees, sales and rentals, and proceeds from parking, advertising, licensing, and other museum activities. Regardless of the extent of such revenue-producing sources, they never produce enough income to cover all the expenses of the institution. But they can close the gap considerably in a high-volume, well-managed situation.

Investment Income

Science centers with large endowments or investment funds receive a substantial portion of their funding from dividends and interest on long- and short-term investments. Few institutions, however, have sizable endowments or investments. As a result, investment income does not represent a major source of revenue for most science centers. Generally, the endowment or investment fund is managed by the finance committee of the board of trustees or its agent—usually a local bank. An endowment fund differs from an investment fund in that the principal cannot be disturbed. Although among the most valuable of revenues, endowment gifts are among the most difficult contributions to obtain.

Admission Fees

Charges for admission are among the most common sources of income for science centers, although some institutions have a free entrance policy. Fees generally are paid for admission to the museum, planetarium, theaters, and/or special exhibitions. The decision to have a free or fee admission frequently is influenced by the acceptance of local or state tax funds, which often carries a stipulation for free admission for all or merely school groups. In some instances, public funds may call for a single free day a week. A free admission policy usually is based on the argument that a museum should serve as many people as possible, particularly if tax funds are used for its support. A change from free to fee admission nearly always results in a decrease in attendance—sometimes as much as 30 to 50 percent. But much of this drop can be recovered with imaginative programming and planning.

Membership Dues

Science centers with membership programs receive a portion of their revenue from annual dues. Such income can be substantial or negligible, depending on the number of members and the cost of serving them. In some cases, membership programs cost more than they produce in revenue. When this occurs, the value of a membership program should be re-examined. Generally, membership programs have objectives beyond membership dues—doing a better job of serving interested persons, improving community participation, developing an annual giving program, and/or encouraging people to remember the science center in their bequests. Most membership programs have a number of member categories—student, individual, family, and one or more supporting categories.

Program Fees

The fees received from educational courses, films, field trips, school services, exhibit maintenance, and other program activities can be extremely helpful. The cost of conducting such programs, however,

often is greater than the income, largely because of poor cost-ac-
counting or the failure to charge sufficient fees for services. At some
science centers, such programs are considered "loss leaders" because they are designed primarily to attract and serve members, to
enrich educational programming, to obtain publicity, and/or to
broaden the community role of the institution rather than to produce
revenues.

A few science centers, such as the Museum of Science in Boston
and the Cleveland Health Education Museum, have developed
school visitation programs for which they are paid by school systems or students. Another approach is used by the Franklin Institute
Science Museum, which charges fees for presenting scientific demonstrations to classes and assemblies in schools. Some science
centers charge for maintaining sponsored exhibits.

Sales and Rentals

Revenues from the sale of souvenirs, books, and food and the rental
of lockers, exhibits, and facilities can be substantial. Nearly every
science center has a sales counter or store and occasionally a separate bookshop. Food services also are common, ranging from
vending machines to full restaurants. Rental income can be helpful,
particularly if the museum produces traveling exhibitions with rental
fees and/or rents its meeting rooms and theaters to community and
conference groups.

Other Proceeds

Some science centers receive income from parking, licensing, advertising, and subscriptions. Such revenues generally are not large;
however, parking is a major source of income at the California
Museum of Science and Industry and the Museum of Science in
Boston. The Los Angeles museum operates the parking lot for the
Coliseum football field, and the Boston museum has a five-level
garage that accommodates 840 cars.

Support

Public and private support are essential to the well-being of all
museums, especially to science and technology centers, which usually are high-cost operations. Tax funds, contributions, grants and

contracts, and/or the contribution of materials and services are needed to balance the budget at all institutions.

The government supports museums in many countries, but in the United States and some other nations museums are largely privately operated institutions. A few notable exceptions are the Franklin Institute Science Museum, which receives annual appropriations from the City of Philadelphia and State of Pennsylvania; the Museum of Science and Industry, which shares in a Chicago Park District property tax for museums on park district land and a State of Illinois appropriation for museums on local public land; the Center of Science and Industry in Columbus, which receives annual support from Franklin County; the Exploratorium, which receives irregular support from the San Francisco hotel/motel tax fund; and the Science Museum of Minnesota in St. Paul, which obtains both local and state support for performing certain services.

The Science Museum of Virginia in Richmond (funded by the state) and the American Museum of Energy and Science in Oak Ridge (supported by the U.S. Department of Energy) are two American centers operated as government agencies. In Toronto, the Ontario Science Centre functions as a branch of the Province of Ontario, and the science centers in France, India, and Singapore are operated by their national governments. The primary danger of government support is government control. Unless a privately operated science center is careful, it may lose its independence and full control of its programming by accepting tax funds.

Contributions

Contributions from private sources—individuals, companies, and foundations—traditionally have been the principal source of support for museums, but they have not kept up with climbing costs. Most private gifts are the result of personal or mail solicitations, frequently as part of annual giving programs, special projects, capital campaigns, bequests, and fund-raising events, such as dinners, auctions, and special shows.

A number of science centers, such as the Detroit Science Center, Omniplex in Oklahoma City, and the Museum of Science and Industry in Chicago, are the result of private philanthropy. Because of the demise of private fortunes, however, such major contributions are rare, and it requires a well-organized development program to obtain smaller amounts from a greater number of people. In some communities, science centers share in "united appeals," such as the Commmunity Fund or United Arts Fund, that solicit contributions on a cooperative basis for a large number of nonprofit organizations in the area. Such united appeals can be helpful, but they also may restrict an institution's own fund-raising activities.

Grants and Contracts

At many museums, grants and/or contracts provide additional support. They usually come from government agencies and private foundations for specific purposes, such as the development of an exhibit, the offering of educational programs, the training of personnel, or for research. A grant differs from a contract in its legal obligations, requiring the best efforts of the recipient but not the delivery of a specific product.

An increasing number of grants have become available in the United States through such federal agencies as the Institute of Museum Services, the National Endowment for the Arts, National Endowment for the Humanities, National Science Foundation, and National Museum Act administered by the Smithsonian Institution. In all cases except the IMS grants, the funds are primarily for projects rather than general operating purposes. Applications must be submitted that summarize proposals and include information about the institution. The ratio of success on grant applications generally is quite low because of the high number of applicants, the failure to tailor proposals to guidelines, and the limited amount of grant funds. Grants and contracts also are not a dependable source of support because of the long review process, extensive paperwork, and the fact that the projects require more money than the funds can provide.

Unfortunately, the United States does not have a "museums program" for science museums comparable to the National Endowment for the Arts program for art museums and the National Endowment for the Humanities program for history museums. The limited federal funding comes primarily through the National Science Foundation's Public Understanding of Science Program for

innovative projects and through the Institute of Museum Services
for operating purposes, both of which are being phased out.

Contributed Materials and Services

Another source of support consists of contributed materials and
services for which values are now being placed in financial statements. Contributed materials range from building supplies to artifacts and include such things as office equipment, exhibits, auction
materials, and science kits. A museum, however, must be careful
not to appraise such material gifts for a donor's tax-deduction purposes. It is the donor's responsibility to set and justify the market
value, usually with the assistance of an independent appraiser.

Contributed services cover such activities as building repairs
made without cost by a friendly contractor, maintenance and security personnel provided by the city, and the services of museum
volunteers and docents. New nonprofit accounting principles require that such services be given a monetary value in financial
statements when a person performs a task that otherwise would
require a paid staff member. Although the accounting entry is a
"wash," it helps to show the full extent of community support and
results in greater financial accountability by the institution.

Fund-Raising Programs

The word "development" frequently is used for fund raising because individual, corporate, and foundation gifts usually must be
"developed" before they materialize; they rarely come in over the
transom. To raise funds successfully, a science and technology center must have an effective organizational plan, specific objectives,
and the appropriate fund-raising tools. Such a vital task cannot be
left to chance nor assigned to a board trustee or staff subordinate
without adequate direction and support.

The museum director has the prime responsibility for raising sufficient funds for the operation of the institution regardless of the
organization, goals, or techniques. Failure to obtain the necessary
funds usually results in severe cutbacks and/or improvement delays
that seriously impair the operations of the institution.

In a small science center, the director may be the entire fund-
raising staff, with assistance provided by members of the board of
trustees. In medium-size and large institutions, one or more persons

may be assigned to assist the director with the fund raising through a development office. On major campaigns, a professional fund-raising firm may be hired to offer guidance and/or to direct the actual solicitation efforts. Generally, professional fund raisers are not used for campaigns of less than $200,000. Payment for such services should be based on time charges rather than as a percentage of the proceeds.

Except for museums that are national in scope and attendance, science centers almost always must depend on local sources for most of their support. It is difficult to convince a national corporation or foundation that it should contribute to an institution unless it has a local plant or the request falls within the specific interests of the organization.

In raising funds, the needs and case for support of an institution must be clearly stated. But the needs frequently are not as important to donor prospects as the opportunities to achieve something they want. "Any person or group which provides significant amounts of money to a museum expects something in return," according to one museum planning guide. "This might be public recognition, a position to influence policy, income tax benefits, improvement in the civic climate for business purposes, or assurance that the public welfare will be enhanced by an investment. It is important to point out all such potential benefits to donors. But it is also important to consider whether the policies favored by significant potential funding sources are compatible with the stated goals of the museum group."[7]

Fund-Raising Goals

Fund-raising goals usually fall in three broad categories: annual support, capital gifts, and deferred giving. In the museum field, two other goals may be added: special project funding and endowment gifts. Each requires a somewhat different approach, although an institution may be seeking support in all areas at the same time.

Annual Support

Annual support is the most common reason for raising funds at science and technology centers. These funds generally are requested for operating support or unrestricted purposes. The "fund raising" may take the form of personal solicitations, a plea through the mail, a budget appropriation request from local or state government, and/or grant application to a foundation or a federal agency. In most cases, annual giving programs are handled by the director, development staff, and/or trustees and are directed at members, friends, corporations, and the local community.

Special Project Funding

With the increase in federal programs for funding special museum projects in the United States has come a comparable rise in fund-raising efforts for such projects as new exhibitions, education programs, minority activities, and programs for senior citizens and the handicapped. Such grant applications—to federal agencies, foundations, and corporations—generally are handled by the director and/or development staff. The project normally is not undertaken unless the funds are awarded because of its impact on the operating budget.

Endowment Gifts

Every museum would like to have endowment gifts, but they are relatively rare. Contributions earmarked for perpetual investment, with the institution receiving only the income for operating purposes, is a difficult concept to sell. Most people would rather give for operating or even capital support, but a sizable endowment fund can produce helpful dividends and interest for any science center.

Capital Gifts

A capital campaign usually is a greatly intensified fund-raising program, requiring a special staff, budget, and organization and involving many community leaders, volunteers, and prospects. Capital gifts are among the most difficult contributions to obtain. Because of the magnitude of the funding and effort, most institutions seldom have more than one such large-scale campaign after their founding. Capital drives for minor additions are more common and frequently are spearheaded by the trustees.

Deferred Giving

Bequests and trusts are forms of deferred or planned giving with great promise. However, most science centers devote little attention to them. This is a fund-raising area that needs cultivation with trustees, members, lawyers, bank trust officers, and the general public. The Museum of Science in Boston has a special pamphlet on bequests, instructing members, friends, and others on how to prepare wills, trusts, and annuities for the benefit of the museum.

Fund-Raising Elements

Raising funds, by necessity, must be a deliberate rather than haphazard process. Certainly, some gifts will be received that were not influenced by a conscious effort on the part of the museum. But the great bulk of support must be sought out and convinced to make a contribution. How well the fund-raising program is organized and implemented usually determines how much an institution receives in contributions. In addition to an effective case for giving and adequate budget and staff, elements of a fund-raising program generally include an organizational plan, leadership, volunteers, prospect lists, solicitations, and fund-raising materials.

Organizational Plan

A science center must consider two aspects of fund-raising organization: how to organize internally to handle fund-raising activities and what organizational approach to use in seeking funds for special purposes. The museum director generally is the keystone of the internal organization, whether he or she does the fund raising alone or with the assistance of a fund-raising counsel, as indicated earlier. The board of trustees and its chairman and/or the finance or development committee also should play important roles in planning and implementing a fund-raising program.

An organizational plan for a campaign requires a much greater effort and somewhat different strategy than day-to-day fund raising. It also involves many more people, a larger budget, and expanded promotional activities. A fund-raising campaign, whether it is for operating support, capital improvements, exhibit program, or other purposes, must have strong leadership, numerous volunteers, identified prospects, appropriate solicitation, and the necessary fund-raising materials.

Leadership

It may be sufficient to rely on a museum's director and development staff for routine fund-raising activities, but it is necessary to involve community leaders to obtain substantial funds for basic support or special campaigns.

This leadership already may be available from members of the board of trustees, but it needs to be organized, directed, and assisted. Many governing boards have a finance or development committee concerned with an institution's financial resources and development. In other cases, it may be more effective to ask a local corporate president, banker, merchant, or other community leader who is not on the governing board to serve as campaign chairman. The chairman should be supported with a well-devised volunteer committee of other prominent community figures and conscientious workers.

Business and industrial leadership is essential if a science center expects to receive major gifts from companies. Many corporate contributions are based on peer solicitations, friendships, and obligations. Such "peer pressure" frequently accounts for the difference between the size of a contribution solicited by the museum staff and one obtained by a corporate executive on the campaign committee. It also brings in the big gifts that should launch and set the pace for a campaign. However, it takes more than a few community leaders to conduct a successful campaign. It requires an effective backup staff, hard-working volunteers, and the necessary tools to do the job.

Volunteers

The foot soldiers in any fund-raising campaign are the volunteers; they are the people who make personal calls, write letters, ring door bells, use the telephone, and/or arrange special events to raise money. A campaign may involve hundreds—and sometimes thousands—of people to canvas an area served by the institution. Volunteer workers usually are divided into broad committee categories, such as business, professional, labor, neighborhoods, women, etc., and then subdivided into smaller units, with five to ten workers headed by a captain. Each volunteer is expected to solicit five to ten prospects, which are frequently selected from a list of prospects or friends. In general, the more volunteers, the more money raised. All campaign workers should attend at least one training session, be assigned a quota or goal, and receive the necessary direction and materials to raise the necessary funds.

Prospect Lists

One of the most essential ingredients and treasured items of any fund-raising effort is the prospect list. Such lists usually are developed over a period of time and are based largely on an institution's past experiences. They must be updated and supplemented continuously by removing from the list names of those who have died or moved out of town, keeping up with changes in addresses and jobs, and adding the names of new prospects.

The size, breakdown, and use of the prospect list will depend on the nature and magnitude of the fund-raising program. It is common to "rate" a list according to the amount or range of contribution that may be expected from a prospect. A list then is divided into major, middle range, and small gift categories.

The prospects should be kept alphabetically in a master card or computer file, generally called a "locator file." From this file other specialized lists can be developed according to a prospect's relationship to the museum (such as trustee, member, donor, etc.); rating or giving potential; or campaign committee assignments. One or more different colored "stop cards" or computer codes may be used to identify prime prospects for special handling when making assignments. So-called "flat lists"—those typed on sheets—also are needed, especially for mailings, selections by volunteer solicitors, and keeping track of assignments.

Solicitation

A science and technology center also must be concerned with many aspects of soliciting funds. For example, it may not be appropriate or necessary to have the elaborate structure of a major campaign for a nominal annual giving solicitation. Nor is it essential to have an army of volunteers when the fund-raising program is based on limited personal solicitations of corporations or grant applications. At a publicly supported museum, the principal fund-raising mechanism may be the budget hearing at city hall or the state capitol. On the other hand, if a new building is planned, a major capital campaign may be the only way to raise the necessary funds.

In conducting a solicitation, a science center should be careful about the timing, quotas, assignments, and amounts requested. Campaigns should not be launched to conflict with annual drives by major fund-raising organizations. All campaigns should have timetables and a system for following up on assignments. In addition to prospects, assignments should include quotas. And in making solicitations, the case must be sold first before indicating the range of contribution being sought. Besides the cause, it sometimes may be appropriate to appeal to other reasons why people give, such as guilt, ego, sympathy, and desire for community betterment or a personal monument.

A science center also must be concerned about creating a favorable giving climate and maintaining the momentum of a campaign. This involves close cooperation with the public relations department as well as with the fund-raising leaders and workers. Giving begets giving, and it is imperative to start with substantial advance gifts and to keep the drive rolling. Once the momentum slows down, it is difficult to rekindle interest and support.

Accurate recordkeeping is essential, as are receipts and acknowledgments for contributors and rewards for campaign workers. The museum staff should handle the paperwork and see that appropriate thank you notes are put in the mail and that campaign workers are recognized in some suitable form, perhaps with certificates and/or at a luncheon or dinner.

Harold J. Seymour pointed out in his fund-raising book that one-third of a campaign goal usually comes from the top ten contributors, another third from the next hundred, and the remaining third from all other gifts.[8] In most cases, he said, about 50 percent of the contributors tend to give the same amount in the previous year, 25 percent give more, 15 percent give less, and 10 percent are new or former donors.[9]

If these ratios are correct, it is obvious that a science center must give its greatest attention to major and past contributors, while attracting or upgrading small and new donors. Unfortunately, the average American company gives less than 1.5 percent of its taxable income to nonprofit causes, even though it can deduct up to 5 percent for tax purposes. But the prospect for a convincing case is always present.

Fund-Raising Materials

Many types of printed, audiovisual, and other materials are needed for fund-raising activities. Some are for recordkeeping and organizational purposes, while others are necessary for personal, mail, and phone solicitations.

Prospect cards, assignment sheets, report forms, pledge cards, form letters, and return envelopes are among the routine printed materials needed for a fund-raising program. A letter, pamphlet or brochure, pledge card, and return envelope usually form the basis of a direct mail appeal. Whenever possible, the letter should be autotyped and personalized rather than printed.

Special Presentations

Special presentation kits tailored to individual prospects sometimes are used as part of personal solicitations. Other times, slide or film presentations are included in the personal appeal or are used at meetings and luncheons for large groups.

Mail Appeals

Nearly every fund-raising effort includes some use of mail appeals. In a major campaign, direct mail may play an important role. Such mailings normally include a printed or autotyped letter, a leaflet or brochure, a pledge card, and a return envelope. In many cases, the pledge form is part of the return envelope.

Grant Applications

Applications for grants usually are submitted to government agencies and foundations by the museum staff. Federal grant requests generally are for special project funding. The practice varies with foundations, which also may provide operating or capital support.

Telephone Solicitations

Although commonly used in social and welfare agency fund campaigns, telephone solicitations seldom are utilized in the museum field. They tend to be inappropriate and ineffective unless the prospects are known to the solicitor, and then a personal visit would be more productive.

Door-to-Door Solicitations

Almost never used by science centers, house-to-house calls seldom produce many donations and nearly always are awkward, even for neighbors.

A number of publications frequently are used in raising funds. They include a leaflet and/or brochure stating the case for giving, a bulletin or newsletter giving the campaign progress periodically, and possibly special pamphlets for specific needs or solicitation areas.

In the audiovisual field, special films, slide shows, and multimedia presentations sometimes are produced for showing at meetings, luncheons, and special presentations and for use on television and in theaters, schools, and plants.

It is highly unusual for an institution to make use of all of these fund-raising materials, mostly because of the cost. Some materials simply are not necessary in the average annual giving or other fund-raising effort.

Fund-Raising Techniques

A science and technology center has an arsenal of fund-raising methods available, such as personal solicitations, door-to-door solicitations, special events, and united appeals.

Personal Solicitations

Personal fund-raising requests are the most common and the most effective. It is difficult to say no to a face-to-face solicitation, particularly when made by a community leader or friend. Personal calls on corporate executives, foundation officials, and acquaintances often can make the difference between a contribution or a turndown and a large or small gift.

Special Events

Special events are another popular and effective technique for raising funds for almost any purpose. Presentation luncheons, benefit dinners, auctions, balls, fashion shows, raffles, theater parties, rummage sales, and other special events produce results when properly organized and promoted. Women's committees frequently are responsible for such activities. At least three science centers—in Portland, Columbus, and Des Moines—hold annual auctions as a means of raising funds.

United Appeals

In some communities, museums receive support from community funds. A relatively new approach is the "united arts fund" concept in which various performing and visual arts organizations, museums, and other cultural groups join together in a joint appeal. Although helpful, such an approach sometimes restricts a museum's ability to raise additional funds. The Omniplex science center in Oklahoma City participates in a united arts drive each year.

Whatever techniques are utilized, raising funds can be a difficult chore. But it is essential for most science centers. Without contributed funds, it usually is not possible to meet operating deficits, provide basic services, expand the facilities, and/or enrich the exhibit and educational offerings.

10

Business Affairs

The operation of a science and technology center requires fiscal and business management whether it is a small institution with limited resources or a large science museum with many facets. Many of the financial principles and practices are the same or similar regardless of the scope or nature of the science center.

The board of trustees is legally responsible for the operation of an institution and its financial stability, but the director is charged with the overall management responsibility and is held accountable by the trustees for having a sound financial operation.

In a small science center, the director may handle all the financial affairs with the help of a secretary and/or bookkeeper and the treasurer from the governing board. A medium-size institution may have a business office headed by a chief accountant or a business manager who reports to the director and works with the treasurer. In a large science museum, the director may find it necessary to have a more fully developed business department with a business manager or administrative director and a supporting staff of clerks, accountants, personnel specialists, and sales people.

Whatever the institutional size or organizational structure, the manager of business affairs has a vital role in the operation of a museum. It is his or her responsibility to prepare the budget, account for receipts and expenses, handle the payroll, oversee the operation of the museum store and/or food services, produce financial reports, manage investments, provide the leadership on fiscal planning, and perform various other duties that affect the financial health of the institution.

The handling of these fiscal functions has become especially important in a period of rising costs and inflation. As a result of increasing demands for greater accountability and standardization of reporting by nonprofit organizations, the museum profession has sought to develop appropriate accounting and financial reporting policies and procedures. With the assistance of grants from the National Endowment for the Arts, the Association of Science-Technology Centers has prepared a helpful *Museum Accounting Guidelines* report and the American Association of Museums has published a *Museum Accounting Handbook* for financial record-keeping and reporting at museums. The *Museum Accounting Guidelines* lists five objectives for developing the recommended policies for preparing financial statements at museums:

1. To bring about more effective accounting and financial reporting practices among museums.
2. To provide more realistic information upon which museum governing boards and management can make policy and management decisions.
3. To develop more uniform practices for comparison and statistical purposes.
4. To meet the increasing need for greater accountability among museums in reporting to the public.
5. To assist auditors in examining and reporting on the financial statements of museums.[1]

In addition, similar efforts have been made by public accountants and others to develop uniform accounting concepts and principles for all nonprofit organizations. The Financial Accounting Standards Board and the American Institute of Certified Public Accountants have issued preliminary reports that will have far-reaching impact on the accounting and financial reporting of all nonprofits when they are finished.[2] The implications of these new nonprofit guides, particularly as they affect capitalization and depreciation, pledges, contributed services and facilities, collections and exhibits, and functional reporting, will be discussed in this chapter.

It is obvious from recent developments that museum financial statements will be receiving greater attention from contributors, public officials, and accountants. Already science centers and other museums are attempting to improve their accounting and reporting systems.

Budgeting

The annual operating budget of a science and technology center has two major functions: it is a plan of action for future activity and a device for monitoring ongoing activity. In other words, it is both a monetary blueprint and control.

For any meaningful budgetary planning, it is essential to know the institution's goals for the coming year. Too often, the goals do not become evident until the budget process is under way. Goals usually are expressed in terms of the programs and activities planned for the next year. In general, these programs and activities are grouped as either program services or support services. Each program and activity should have an objective and cost estimate.

Regardless of the approach, six basic steps are required to complete the budget preparation and review process, as explained in the *Museum Accounting Handbook*:

1. Prepare a list of program objectives or goals for the following years. These objectives serve as performance measures against which the actual activity will be evaluated.
2. Determine the resources required to attain the goals and estimate the cost of these resources.
3. Estimate the cost of each program.
4. Estimate the income expected from each source, giving careful consideration to economic conditions.
5. Compare the total expected income with the estimated cost of achieving the listed objectives or goals. The expense figures must be adjusted until they produce an excess or deficit figure that is acceptable to the organization. This is accomplished through a review of program priorities.
6. Present the final proposed budget to the appropriate body for a thorough review and ratification.[3]

Once the annual budget is approved, it is necessary to break down the budget into periods corresponding to the periods of interim financial statements, which normally are prepared monthly or quarterly. It is a mistake, however, to divide the annual budget into equal amounts, since there are fluctuations in seasonal energy costs, attendance patterns, and exhibition schedules. An attempt should be made to apportion budget items monthly or quarterly according to a projection or a profile of the prior year's actual results.

Interim financial statements must be prepared promptly and accurately to be of value in fiscal management. They should show the totals, budget, and variances for the period and the estimated totals and budget for the year. If it becomes apparent that expenses are exceeding the budget or that the income will not be as high as anticipated, it is imperative that the appropriate body—usually the museum director or the governing board's finance committee—acts promptly to modify operations. The best-prepared budget is of little value unless the museum director and/or the board is willing to take whatever steps are necessary to make adjustments.

Accounting System

Many museum people confuse the terms "bookkeeping" and "accounting." Actually, bookkeeping is the process of recording financial information within an accounting system. Accounting is the system for gathering, summarizing, and reporting financial information; it is the framework for budgets, financial planning, and internal and external reports and the heart of financial recordkeeping, reporting, and management.

The *Museum Accounting Guidelines* report recommends that museums follow generally accepted accounting principles. Some of these principles are quite specific, while others may offer options. But the principles should be the foundation of any accounting system. They also are the basis of an independent auditor's report.

Selecting an Accounting System

How should a science and technology center present a distillation of thousands of financial transactions? What should be the basis for maintaining records or accounting funds? These and other questions are influenced by accounting principles and institutional preferences.

Maintaining Records

There are three basic choices for maintaining financial records: cash, modified cash, and full accrual accounting. In cash accounting revenue and expense transactions are recorded only upon receipt or disbursement of cash. Small museums using a bookkeeper without formal accounting training will frequently use this procedure. An independent auditor, however, cannot render an opinion that the financial statements are prepared in accordance with generally accepted accounting principles under this system.

Modified cash accounting is similar to the cash approach for recordkeeping, but worksheet adjustments to the books are made at the end of the accounting period to reflect certain unrecorded noncash activities. Most adjustments relate to accounts payable, unpaid salaries, vacations due, museum store and food inventories, prepaid expenses, and accounts receivable.

Financial statements based on full accrual accounting more accurately present financial information, primarily because they are not usually affected significantly by the timing of cash receipts and disbursements, which is the case with the cash method. Revenues are reported when earned and expenses when incurred; those applicable to future periods are deferred. Medium-size and large science centers with high volumes of receivables and payables, significant amounts of prepaid expenses and inventories, and a large number of financial transactions generally find accrual accounting to be more advantageous and less cumbersome than the cash or modified cash approaches.

Accounting for Funds

A "fund" is a separate account created to insure the limitations and restrictions placed on the use of the resources as specified by donors, the governing board, and/or outside regulations or restrictions. There are three methods used to account for funds: one fund accounting, simplified fund accounting, and full fund accounting. A one fund accounting method is used when a museum operates with only unrestricted assets, funds, and grants. Simplified fund accounting is an adaptation of the one fund approach. When a single gift or grant is made to an institution for a specific purpose, it is necessary to segregate this asset from the unrestricted or operating funds. Thus a separate restricted fund can be established without imposing all the complications of a full fund accounting system.

The preferred system, however, is full fund accounting, under which a museum's assets are separated into sets of self-contained and self-balancing accounts, each having its own assets, liabilities, and fund balance and functioning almost as separate organizations.

There are four basic fund groups in full fund accounting: unrestricted or operating fund, current restricted fund, plant or capital fund, and endowment fund. An unrestricted fund provides the basic operating support for a museum; assets placed in this fund have no restrictions placed on them by external authority. Sometimes an unrestricted fund contains a board-designated investment fund, or quasi-endowment fund. These amounts, which often are accounted and reported separately, have been set aside for investment by the board on its own initiative.

A current restricted fund contains assets that are expendable for operating purposes but are restricted by donors and grantors for specific uses. Assets restricted to endowment or physical plant use are recorded in this fund only when there is not a separate plant fund.

The fixed assets of a museum usually are kept in a plant fund, regardless of whether they have been restricted by outside authority. The basis for such segregation is that these assets are committed as permanent assets and are not available for the day-to-day operation of the museum. If an institution prefers, however, it can record fixed assets in the current unrestricted fund. Contributions restricted for physical plant repairs or improvements are recorded in the plant fund when it exists and in the current unrestricted fund when it does not.

An endowment fund contains assets given to the museum on the condition that the principal of the gift be kept intact and that only the income produced by investment be used to meet the museum's needs. The income may be restricted or unrestricted, depending upon the terms of the gift. Endowment funds are different from board-designated investment funds that are recorded as part of unrestricted funds.

Functional Accounting

A relatively new concept in accounting is the reporting of expenses along functional or programmatic lines rather than the more traditional basis showing amounts for salaries, materials, travel, equipment, etc. Functional accounting, as it is called, can be extremely helpful in determining the "real" cost of an activity based on direct expenses, the time spent on the activity, and a share of the facility expenses, depreciation, and general overhead. Such a system requires better recordkeeping and allocation of certain expenses. For example, if a person has several functions, it would be necessary to determine how much time is spent on each program.

Functional accounting makes it easier to evaluate museum activities and to compare costs of different programs at the museum and with those of other institutions. But functional accounting also is more time-consuming and complicated than standard accounting.

Museum activities are usually classifed as either program services or supporting services in functional accounting. Auxiliary activities, depending upon their nature, may be separated into a third category. Typical program services for science and technology centers include collections, exhibits, education programs, membership activities, and community services. Supporting services generally are classified only as management and general expenses and as fund-raising costs, but a finer breakdown can be applied if useful. With the increasing emphasis on accountability, functional accounting also makes it possible to relate the total fund-raising cost to the total support received.

Accounting Considerations

The establishment of an accounting system involves various other considerations, such as how to handle capitalization and depreciation; cash contributions and pledges; contributed materials, services, and facilities; investments; auxiliary activities; and related organizations.

Capitalization and Depreciation

One of the most controversial areas of museum accounting is the subject of capitalization and depreciation. Generally accepted accounting principles for most types of organizations require that land, buildings, equipment, furniture, vehicles, and other fixed assets be carried at cost and expensed over their useful lives with an annual depreciation charge. However, many museums write off assets when purchased as current expenses.

In all likelihood, the accounting profession will require museums to capitalize and depreciate fixed assets in the future. It is argued that fixed assets are a major asset for which a governing board is accountable and that failure to reflect these assets on a balance sheet is misleading. Capitalization also presents a more comparable statement of activity from period to period, without the distortions of a large acquisition in any period.

The most difficult aspect of capitalization is the requirement for retroactive recording of fixed assets. If the historical cost of assets still in use cannot be reconstructed, other reasonable bases may be acceptable for the initial recording, such as cost-adjusted appraisals, insurance appraisals, replacement costs, or property tax appraisals adjusted for the market. This latter approach applies only to the initial recording at the effective date and only when historical cost records are not available. All subsequent additions must be recorded at cost, or at fair value if donated.

Depreciation accounting is an extension of capitalization. It is a method of allocating or spreading the cost of an asset over its useful life (less salvage). It is not intended to be either a valuation process or a means of providing replacement funds. The objective is to more accurately reflect the cost of operations for a period so that a donor, member, public agency, or anyone else can better evaluate the financial results. According to accountants, since buildings and equipment do wear out, depreciation must be considered if one is trying to measure the full cost.[4]

Theoretically, fixed assets would include public land and buildings occupied by museums, collections, and exhibits. However, most museums do not capitalize or depreciate collections and exhibits or public facilities they do not own. But changes are likely to occur in these areas as more uniform accounting principles evolve for nonprofits.

Even though the land or building provided by a city government, park district, or other government agency normally is not capitalized, the improvements made with museum funds to buildings on public land should be reflected in the balance sheet.

On the question of museum collections, independent accountants have argued that they should be capitalized for stewardship and reporting reasons. However, because of the strong objections by museums, accountants have concluded that it is desirable but not necessary. Museums are opposed to the capitalization of collections because of the difficulty in determining the value of artifacts. Where records or values do exist for collections, the accounting profession encourages capitalization at cost if purchased or at fair market value if donated. Rather than incorporate the value in a financial statement, some institutions disclose the information in a footnote.

Constructed exhibits with a limited life are a different story. Greater pressure will be applied by auditors to amortize such "exhaustible" exhibits without objects of intrinsic value over their useful lives. When capitalized, such exhibits should be valued on the basis of their original cost or, when donated, on the basis of their fair market value at the date of contribution. The capitalization and depreciation of constructed exhibits could have considerable influence on the financial statements of science and technology centers in the future.

Cash Contributions and Pledges

Nearly every museum receives cash contributions and pledges. Contributions are either unrestricted or restricted. When unrestricted, they usually go into the operating fund and can be used for any purpose. Restricted gifts are for specific purposes, such as the endowment, a capital program, or a specific operating project.

Most museums do not include pledges in their financial statements until the cash is received, but the accounting profession believes that pledges should be recorded as a receivable on the balance sheet less some reasonable allowance for uncollectible pledges based on past experience. Although few museums attempt to enforce the legal collection of pledges, the recording of pledges gives recognition to the fact that pledges are assets.

Cash contributions and pledges should be recorded as support as received unless all or portions are specified by the donor for use in future periods. When the latter occurs, such future portions should be recorded as deferred support in the balance sheet of the appropriate fund and as support in the year(s) in which they may be used.

Contributed Materials, Services, and Facilities

In general, contributed materials, services, and facilities are not recognized in the financial statements of museums. The accounting profession, however, argues that such recordings are essential if the institution is to properly account for all transactions and fulfill its stewardship role.

Donated materials of any significance should be recorded at their fair value when received, assuming that there is a clearly measurable basis for the value, such as appraisals, price lists, or a market quotation adjusted for deterioration or obsolescence. When a valuation cannot be substantiated or is of little consequence, it is not necessary to recognize the contribution in financial statements. When donated materials are used in providing a museum's services, the cost of such materials should be based on the value previously recorded for the contribution.

Volunteer and other contributed services are most helpful to museums but rarely are reflected in financial statements. The accountants say they should be recorded when they are significant and replace a paid position. The *Museum Accounting Guidelines* report states that the monetary value of services should be recorded in financial statements when they are significant and all of the following circumstances exist:

1. The services performed are a normal part of the museum's program or supporting services and otherwise would be performed by salaried personnel.
2. The museum exercises control over the duties of the contributors of the services.
3. The museum has a clearly measurable basis for the value placed on the services.[5]

Among the types of contributed services not normally recorded in financial statements are the occasional efforts of volunteer workers who do not replace paid employees, the nonadministrative workers in a fund-raising campaign, and unsupervised personnel engaged in research or training activities without pay.

If contributed services are recognized in financial statements, then the statements, according to the accountants, should indicate the methods used in evaluating, recording, and reporting contributed services. They also should distinguish between those services for which values were recorded and those that were not included in the financial statement.

Some museums benefit from being housed in municipally owned buildings on public lands and/or from receiving certain security, maintenance, and perhaps energy services from local government. Although most museums do not include such assistance in their financial services, the accounting profession would like to see the fair value of the use of such facilities and services reflected in the financial statements as support.

Investments

Most museums have two types of investment portfolios. One is board designated and unrestricted and the other is donated and restricted. Unrestricted funds are invested to earn income until the funds are needed for operating purposes. The second type includes investment of endowments and other restricted funds.

The value of investments can be carried in financial statements either at cost or at market value; the latter is preferred. Whichever valuation basis is used, it is recommended that the other value be shown either parenthetically or in a footnote.

The cost of investments purchased by museums should include any brokerage fees, taxes, and other charges directly applicable to the purchase. Contributed securities should be recorded at their fair market value on the date of the gift.

Investments are considered short term or temporary when there is a ready market for converting such securities into cash and management intends to sell them to meet cash requirements. Long-term or permanent investments include stocks, bonds, mortgages, and similar debt instruments and miscellaneous investments such as real estate and interest in trusts and estates.

The accounting for income on investments, including unrealized gains and losses, should be on a fund basis, unless terms of the restricted account permits transfer to unrestricted or other funds.

Some museums have adopted a "fixed rate of return" method of determining the amount of endowment income available to support operations each year. Under such a plan, a predetermined return on endowment funds is set aside to pay for operating expenses. This return is frequently based on the average market value of funds for a preceding period of at least three years.

Another investment technique is "pooling." To obtain greater flexibility, some museums combine investments of various funds. Because the realized and unrealized gains or losses and the income are not identified with specific funds, it is vital that these returns are allocated equitably. To accomplish this, a "market value method" usually is applied and each fund is given an assigned number of units based on the relationship of the market value of all investments at the time they are pooled. These pooled assets are valued periodically and new unit values assigned. The investment pool income is allocated on the basis of the number of units held by each participating fund.

Auxiliary Activities

Although museums are nonprofit institutions, they frequently operate profit-making activities, such as souvenir and book shops, restaurants, theaters, tours, licensing, and publishing ventures. These activities should be included in a museum's financial statement. They also must be related to a museum's purpose or be subjected to taxation.

An auxiliary activity sometimes is operated as a separate corporation for which the museum is fiscally responsible, such as a research foundation or museum press. In such instances, its finances should be reported in the museum's financial statements and perhaps accompanied by a separate disclosure of financial operations.

The total revenue and expenses of auxiliary activities should be reported in financial statements. Some museums also show the net results of operations either parenthetically or in footnotes.

Related Organizations

The financial handling of related organizations, such as volunteer, student, and fund-raising groups, will depend on their structure and whether they are under the control of the museum. If a related organization has its own charter, tax-exempt status, and governing board, it is considered independent and should not be a part of a museum's financial statement. If the organization actually is controlled and its resources are handled by the museum, then it should be included in the museum statement.

Financial Reporting

All the accounting records of a museum are maintained to report how institutional resources have been used and to disclose the financial health of the museum. At least two financial statements should be used to report on the financial status of the organization: the balance sheet and the statement of activity. Two optional statements that also are helpful are the statement of changes in financial position and the statement of functional expenses.

Whatever financial statements are utilized, it should be remembered that the purpose of any statement is to communicate. Malvern J. Gross, Jr., a leader in the nonprofit accounting field and a partner at Price Waterhouse & Company, believes a good way to determine whether a financial statement is accomplishing this objective is the so-called "grandmother test." Institutions should ask themselves whether the financial statements can be understood by "any interested 'grandmother' of average intelligence who is willing to take some time to study them."[6]

Balance Sheet

Sometimes called a statement of financial position, the balance sheet shows the assets, liabilities, and fund balances of an organization at a given date. In effect, it gives the financial health of a museum at any particular moment.

Statement of Activity

All the financial activity of an institution for a fiscal year or other specific period is presented in a statement of activity. It provides information for each fund and a total of all funds regarding the revenue, support, expenses, and changes in fund balances for the period. This financial statement also is known as a statement of income and expenses; statement of financial operations; or statement of revenue, expenses, and changes in fund balances.

Statement of Changes in Financial Position

This optional statement shows all the changes in a museum's assets, liabilities, and deferred accounts over the period covered in the statement of activity.

Statement of Functional Expenses

When functional accounting is used, this optional statement shows a breakdown of museum program expenses along functional lines.

Bookkeeping Process

The process of making records of financial transactions is called bookkeeping. It is the mechanical aspect of accounting. Basically, there are three steps to bookkeeping: documenting, recording, and summarizing. The bookkeeping procedures at various institutions will differ somewhat, based on the size, needs, and preferences of the organization, but they normally include internal controls to safeguard the assets.

Bookkeeping Concepts

The prime objective of any bookkeeping process is to produce accurate and useful information for financial reporting purposes. Without a well-designed bookkeeping system, financial statements simply cannot be reliable. The system should encompass methods for documenting, recording, and summarizing all the financial transactions of a science and technology center.

Documenting

Every financial transaction should have a document showing the amount, the date, and the people involved. Such documents provide the raw data for accounting records. Documents can take almost any form, such as cancelled checks, loan notes, bank deposit receipts, contracts, and other such financial instruments.

Recording

Once transactions are documented, they should be listed chronologically in a journal. This process is called "recording" or "journalizing." A journal can take various forms, but the most common is the checkbook, also known as the bank book or check register. The purpose of any journal is to give a record of each transaction and a running balance in the account.

Summarizing

Every financial transaction has at least two effects, and these effects must balance out. The assets of a museum equal the liabilities and fund balance. Thus, any transaction that changes the assets also must change the liabilities and fund balance, and vice versa. The process of showing the effects of each transaction is called "summarizing" or "posting."

A bookkeeping system must segregate all transactions into categories for financial statements. Each category used to gather financial information is called an account, and each account is written on a separate page in a ledger. Therefore, any transaction recorded in the journal results in an increase or decrease in two or more account pages in the ledger. The balance recorded on any of these pages indicates the cumulative effect of all transactions on that account.

The left side of a ledger page is called the debit side and the right side is the credit side. A debit entry shows an increase, and a credit entry indicates a decrease. Asset accounts normally accumulate balances on the debit side of the page, and liability and fund balance accounts show balances on the credit side.

Increases and decreases in fund balance, however, are not entered directly in the fund balance account. Rather, fund balance increases (income) are accumulated in one set of accounts and decreases in fund balance (expense) are placed in another. The balances in the income and expense accounts is zero at the beginning of the year; at the end of the year, the balances show the total increases and decreases in fund balance.

Bookkeeping Procedures

The bookkeeping procedures used to handle financial transactions will depend to a considerable degree on whether cash-basis or accrual-basis accounting is adopted. Many small museums use cash or modified cash-basis accounting, both of which are relatively simple. Medium-size and large science centers with greater needs usually utilize more complex accrual accounting methods.

Cash-Basis Bookkeeping

Under cash accounting, revenue and expense transactions are recorded only upon receipt or disbursement of cash. The bookkeeping for a modified cash-basis system is the same, but it is necessary to make worksheet adjustments at the end of an accounting period to reflect certain únrecorded noncash activities.

In an extremely small and simple science center, a checkbook may be the only bookkeeping form needed. However, it is necessary to record each check and deposit carefully, giving an explanation of the source or purpose. This should be followed by grouping similar transactions together and preparing a statement from the group totals.

The basic bookkeeping tools of a cash-basis system are journals and registers for recording the details of transactions and a ledger for summarizing information from the journals and registers. A cash receipts journal, a cash disbursement journal, a general journal, and a payroll register should be used for posting information in a general ledger. As the names imply, the respective journals and register record cash received, cash spent, entries not involving cash, and employees' earnings and withholdings. The general ledger is a summary record of the financial activity in each of the museum's accounts for a specified period, usually a month.

A number of bookkeeping actions take place in the ledger posting process in the cash-basis system. They include checking a "trial balance," "closing" the books, and making "adjustments." To uncover errors before drawing up a statement, the debits and credits are checked in a sample or trial balance. The closing of books occurs when the income and expense are entered in the fund balance at the end of the year. Adjustments for unrecorded noncash activities also are made at this point in a modified cash-basis system.

A key element in the process is the chart of accounts. Under this system each of an institution's assets, liabilities, fund balances, revenues, and expenses is assigned a number and is listed in a manner that facilitates the preparation of financial statement.

Accrual-Basis Bookkeeping

Museums with large volumes of transactions—particularly accounts payable—find the accrual-basis system to be more practical than either the cash-basis or modified cash-basis bookkeeping methods. Under this approach, the records and financial statements reflect all transactions and economic events for a period.

The basic records used in accrual accounting are similar to those of the cash-basis system, but they are more numerous, more complicated, and more time-consuming, and thus costlier. They include a support and revenue journal, a cash receipts journal, a voucher register, a cash disbursements journal, a general journal, a payroll register, and a general ledger. Most of these records were discussed previously. The support and revenue journal is used to record revenue from billings to customers and written pledges of support, and the voucher register is used to control invoices and record asset purchases at the time the vendor invoices are received.

Accrual-basis bookkeeping attempts to integrate all financial events into the accounting system and to apportion certain expenses that are a function of the passage of time, such as rent, insurance, and interest. In deciding which bookkeeping system to use, a science center should determine what information is needed and then choose a method that produces the information with the least investment of time and money.

In his helpful booklet, *Up Your Accountability,* Paul Bennett states that "It is essential to the public trust of nonprofit organizations that proper records be kept to allow public scrutiny and to aid management in improving the organization's efficiency. But it is a violation of that trust to spend, on unnecessary bookkeeping, money that has been contributed for the benefit of the community."[7]

Internal Controls

Regardless of the bookkeeping method, every museum must safeguard the assets of the institution through internal controls. Too often a museum will not utilize all the available controls for fear of offending employees. However, internal controls are designed as much to prevent accidental losses as thefts, and they even can be helpful to employees by protecting them from suspicion.

Among the common internal controls are check signatures, bank reconciliations, petty cash receipts, cash receipts double-handling, purchase orders, and equipment inventory control. The double-signature check is an effective internal control, unless the counter-signature is automatic. Bank statements must be reconciled with the museum's records to make certain all deposits have been recorded and to prevent unauthorized charges. The availability of a petty cash fund is helpful, but receipts should be required for expenses.

For cash receipts, it usually is a good idea to separate the cash-receiving from the recordkeeping. The division of functions and the cross-checking make losses and inaccuracies difficult. The use of purchase orders in dealing with supplies also has benefits, especially when bids or estimated prices are obtained in advance. Equipment losses also can be reduced substantially by assigning someone to inventory control and using a checkout register.

Other Aspects of Financial Management

Financial management involves a number of areas that deserve special mention because of their importance and/or complexity.

They include the handling of payroll, insurance, retirement, sales activities, and taxes.

Payroll

It is extremely important for an institution to keep proper payroll records. Basically, the payroll function includes paying employees the net amount earned, withholding taxes for payment to the government, withholding amounts for various optional purposes, and paying the employer's payroll taxes.

In the United States, the museum serves as an instrument for the payment of employees' income and Social Security taxes. Other payroll deductions that frequently occur are for life and health insurance, retirement plans, and contributions. Because of the increasing ramifications of the payroll system and growing requirements of fair employment, affirmative action, retirement, and unemployment compensation legislation, some institutions have hired personnel managers to oversee payroll and other employee activities.

Insurance

Most museums have an assortment of insurance policies that they must administer. Some cover the institution, such as property, general liability, automobile, and worker's compensation, while others pertain to employees, such as life, health, and long-term disability insurance. An increasing number of museums also are providing fiduciary insurance to trustees as a result of recent legal rulings that make them personally liable for institutional actions.

Because of the value of collections and the cost of insurance, some museums prefer to be self-insured in that area. However, many institutions have fine arts floater insurance policies that cover both the collections and exhibits.

The museum contribution to employee insurance plans varies from merely making available life and/or health coverage to interested employees to paying a portion or all of the cost for such insurance. Most science centers pay at least part of the cost of insurance for employees and usually make it available for dependent coverage. Among the new insurance services are the dental and health maintenance organization (HMO) programs on preventive care.

Retirement

Nearly all museums have some form of retirement plan for their employees. Such pension plans are either contributory or fully funded by the institution. They usually provide for ten-year vesting, retirement at sixty-five, and optional early retirement at fifty-five with reduced benefits.

With the passage of the Employee Retirement Income Security Act (ERISA), retirement funds held by a museum on behalf of its employees no longer belong to the institution and normally are not included among its assets in financial statements. Most retirement plans are not fully funded, but regular institutional payments must be made to the bank or other financial institutions charged with investing the funds.

Sales Activities

The management of souvenir and book shops, vending machines and restaurants, and frequently theaters, tours, and other money-producing activities, such as licensing and publishing, are the responsibility of the person charged with fiscal and business affairs. The proceeds from these sales activities must be included in an institution's financial statements.

Museum stores and food services are making an increasing contribution to the support of institutions. Some science centers have sought to expand their markets by direct mail. Among the institutions that mail sales catalogs to members, friends, and mailing lists are the Franklin Institute Science Museum in Philadelphia, Lawrence Hall of Science in Berkeley, and the Rochester Museum and Science Center.

Licensing agreements for merchandise reproduced from museum collections also are being used more widely, but most science centers do not have the valuable artifacts from which to make salable replicas. More frequently they are found at art, history, and natural history museums.

Taxes

Although most museums are nonprofit organizations, they still must pay taxes on unrelated business income and certain other activities. Fortunately, admission fees, dividends, restaurant income, gift sales, and other such revenues usually are considered related to the purpose of an institution and generally are not taxable on that basis.

The problem becomes more difficult when a museum store sells objects that are not related to an institution's mission or are for the convenience of its visitors. Among the sensitive areas are restaurants serving the neighborhood, theater showings of commercial films, and renting facilities for community use. In some cases, it may be necessary to pay taxes on that portion of the income that is considered unrelated. Museums already are paying Social Security, unemployment compensation, utilities, sales, and other taxes related to their operations.

A science center's nonprofit status is more significant in another respect: it permits tax-deductible contributions. Such donations are the lifeblood of many American institutions. Most countries do not have such a provision and it severely hinders the raising of funds from private sources.

11

The Building

Old or New Building?

A science and technology center can operate efficiently from either an adapted existing building or a new structure designed specifically for its use. But that should not imply that any building is suitable for a science center.

When the founding group has a choice, it usually will opt for constructing a new building over renovating an old one. However, because of limited funds, relatively few science centers have that choice. It also may be more practical for an institution with the necessary support to use a two-step approach, starting and operating in an existing structure to better determine its needs before committing itself to designing a new facility.

Whether the building is old or new, it must be flexible, efficient, and relatively easy and economical to maintain. It is not necessary to have a monumental structure or even an aesthetically pleasing building. It is how the science center functions rather than how it looks that makes the difference.

"Most of today's museums reflect their architects' stylistic leaning at least as much as any genuine attempt to deal afresh with the museum problem," according to Paul Goldberger, architecture critic for the *New York Times.* "And no more than a handful of new museums have earned the unanimous plaudits of museum directors around the country—a group whose members have a certain tendency to ignore problems in their own museum buildings, but who are notably quick to point out shortcomings in others."[1]

Before deciding on a site and building, it is essential for the founders to better define the purpose, nature, and needs of the institution. Is it going to be a participatory or collection-oriented museum? Will it have large and heavy machinery, or will it emphasize ecology and need nature trails? Who will it serve? Will it offer courses or conduct research and need classrooms and laboratories? Should it include a library, restaurant, or school group receiving area? What about security, parking, school lunches, and meeting rooms?

These are the kinds of questions that must be considered before a decision can be reached intelligently on the location and type of building needed. To assist in making the decision, a building committee frequently is appointed to study the institutional needs, to explore the possible costs and financial support, to investigate the choices in buildings, and to make recommendations.

The site of a science center may be as crucial as the building itself. In general, a central location is preferred over one in the outskirts of a city or in the suburbs because of its accessibility. It also is easier to find a suitable structure for renovation or rent in the city. But there may be problems with parking, crime, or insufficient or expensive land on which to build or expand. The nature of the science center's activities also may require open spaces, wooded areas, or other special features generally not found in the heart of the city.

All of these questions may become academic when someone offers to donate or make available an old mansion, school, railway station, courthouse, factory, warehouse, or fair building. But it is necessary to resist the temptation to accept such gifts until their advantages and drawbacks are weighed. A free building may impose a huge operating burden on the institution or distort the science center's activities because of its location or physical limitations.

The adaptation of an existing structure frequently is the best solution to the need for a physical plant. However, the decision should not be made without considering all the needs and implications.

A number of science and technology centers are located in structures that were built for world's fairs and expositions. Some of the conversions have been successful while others have presented difficulties. The Museum of Science and Industry in Chicago is housed in the old Palace of Fine Arts from the 1893 World's Columbian Exposition. Originally a temporary structure, the building was rebuilt and modernized in the 1930s to house the museum on a permanent basis. It has proved to be a comfortable, flexible, and efficient home for the Chicago museum.

The New York Hall of Science inherited a structure from the New York World's Fair of 1964–1965. Although attractive architecturally, the building's configuration and lack of floor space are a handicap to the effective operation of the science center.

Two other American science centers are located in former fair buildings. The Pacific Science Center in Seattle has one of the most beautiful settings. Five interlocking buildings enclose a courtyard of reflecting pools and fountains, but the separate structures sometimes present great difficulties. The Exploratorium in San Francisco is located in a hangarlike building from the 1915 Panama-Pacific International Exhibition. Although the structure provides ample flexible space, it lacks the cohesiveness and finish of an inviting environment.

Abandoned railway stations are finding new uses in many communities. It is possible to convert them into productive science centers, but it is costly. The Science Museum of Virginia has taken the first steps in renovating the Richmond station, and the Kansas City Museum of History and Science has plans for converting Union Station into a physical and life sciences division.

Three of the largest children's museums—in Brooklyn, Boston, and Indianapolis—were founded and thrived in old houses before moving to new quarters. The Cleveland Health Education Museum renovated and added to a mansion for a satisfactory facility. As a rule, however, it is more difficult to operate a science center in a house because of its need for larger spaces and heavier floor loadings.

Various other types of converted spaces are used by science and technology centers. For example, the California Museum of Science and Industry in Los Angeles uses a renovated agricultural exhibition hall and a retired armory as part of its facilities; the Center of Science and Industry in Columbus converted an old auditorium into a multifloored science center with a new facade; the Detroit Science

Center was operated in a large garage before moving into a new building; and the Impressions 5 science center in Lansing was housed in a warehouse in an industrial park before moving into a renovated downtown building.

New structures for science centers come about in different ways. A community fund-raising campaign is most common, but a new building can be the result of the generosity of a local philanthropist, industrial support, or the decision of an educational institution or distant government agency. Sometimes, all the facilities are completed at one time; other times, the construction continues piecemeal over a number of years (usually because of the lack of sufficient funds or programming). The Museum of Science in Boston, for example, has been involved in raising capital funds for more than a quarter of a century. It first began its core unit in 1949, and later added exhibition halls (1951, 1961, 1972), an auditorium and planetarium (1958), and a parking garage (1968–1972).

The Oregon Museum of Science and Industry in Portland made use of a community "barn-raising" in 1958 to erect the first phase of its building. Materials and labor were contributed to build the structure over a weekend. Since then, a planetarium and other facilities have been added.

Major gifts from philanthropists enabled the Buhl Planetarium and Institute of Popular Science in Pittsburgh, the Detroit Science Center, and the Omniplex science center in Oklahoma City to construct buildings and the Science Museum of Minnesota in St. Paul to make a substantial addition. It was industrial support that made possible the buildings housing the Evoluon in Eindhoven, the Swiss Transport Museum in Lucerne, and the science centers in Japan.

The DeKalb County Board of Education and federal grant funds made possible the Fernbank Science Center facility in Atlanta, while the University of California at Berkeley financed the building housing the Lawrence Hall of Science.

Government agencies also are involved in funding science center structures. In the United States, the U.S. Department of Energy built the American Museum of Science and Energy building in Oak Ridge, Tennessee. The buildings housing the Ontario Science Centre in Toronto and the science centers in India, Mexico, Singapore, and Hong Kong also were the result of government funds.

Building Requirements

The experiences of most science centers clearly demonstrate the importance of looking beyond the founding period to when an institution matures and/or expands its activities. Circumstances may make expediency the necessary course, but the founders, trustees, and director should keep the future in mind. In most instances, science centers still occupy the same buildings in which they were founded. Thus, they must live with their mistakes for an extended period.

Certain basic physical requirements apply to all science centers. It is desirable, whenever possible, to develop a master plan, even though its execution may be delayed. Such a plan involves the close cooperation of an exhibit designer and an architect with the founding group (or with the governing board, director, and staff of an operating institution).

"The architectural maxim that 'form must follow function' is the paramount consideration in planning a museum building," G. Ellis Burcaw points out in a guide to the museum field. "The building must be created for or adapted to the needs of the museum operation. Too often the converse is the case; the functioning of the museum often must be adapted to fit the building."[2]

A number of science and technology center buildings—for example, those housing the Maryland Science Center and the American Museum of Science and Energy—were constructed before it was determined what exhibits would go into the structures. The result was a loss in efficient utilization of space and effectiveness in the presentation of the exhibits.

A well-developed master plan is invaluable in formulating exhibit, program, and architectural concepts. It also can be extremely useful in selling the science center idea, particularly when accompanied with sketches, models, and/or slides.

In determining whether an old building should be converted or a new building constructed, the planning group must go beyond the initial cost of renovation or construction. It must consider the cost of operating the structure after occupancy—including the energy bill, personnel requirements, security, etc.—and the building's suitability for the proposed exhibits and programs. Consideration also should be given to the need for adequate parking, the handling of school groups, the proximity to major thoroughfares, the possibility of expansion in the future, and the ability to raise the necessary funds. Selecting a building for a science center is a complex matter that involves more than cost.

Whether it is an old or new building, there are certain physical needs that should receive attention. As Daniel Traverso has pointed out in a museum planning guide, the ceilings should be twelve to fifteen feet in height; the doors and elevators should be large enough to move oversized objects; floors should be able to support heavy loads, ample storage, service, and work space should be provided; and programming should consider efficiency and expansion.[3] There also should be adequate electrical power, few windows, great flexibility in the exhibit areas, and a horizontal traffic flow (multifloor plans should be avoided if possible).

The typical small science center has from 25,000 to 50,000 square feet of space; most medium-size institutions are in the 50,000-to-100,000 range; and large science centers have from 100,000 to 600,000 square feet. But it is possible to have an effective science center with as little as 10,000 square feet.

Physical Plant Care

Building maintenance is an essential and costly aspect of any science center's operation. How this function is organized and carried out can have profound implications for the budget and the satisfaction of the visiting public.

In a small institution, a single janitor-handyman may be sufficient to clean, heat, repair, and handle other phases of building maintenance. In medium-size and large science centers, caring for the building may require an engineer as well as janitors, stationary engineers, electricians, carpenters, plumbers, and groundskeepers. The person in charge of the physical plant generally is known as the building engineer or manager.

In many cases, building maintenance is the responsibility of the business or administrative department. Larger institutions sometimes have a separate building maintenance department. However organized, care of the building should receive the highest attention, particularly with the rising cost of energy, the increasing number of liability suits, and growing requirements for the handicapped.

Typically, the building maintenance function has involved the cleaning, heating, cooling, and repairing of the building. Roof repairs, plumbing, electrical work, grounds care, painting, furniture repair, and operating the boiler room and air conditioning system are among the principal responsibilities. Some of these tasks must be performed daily, while others are carried out seasonally or periodically. In some instances, it may be necessary to utilize outside contractors for the work.

It is essential for a building to be kept clean, safe, and in good repair. Every effort should be made to remove debris and graffiti and to repair inoperative and vandalized materials. Such conditions merely encourage others to contribute to the problem.

A building maintenance department also may be charged with security and safety, the planning of capital improvements, conservation of energy, and improvement of access for the handicapped. More often, all or part of these responsibilities are shared with or assigned to others, such as the director, business manager, and/or an outside architect or contractor.

Whether it is planning or caring for a building, every science and technology center should devote as much attention to the physical plant as it does to exhibits and programs. A poorly planned or dirty building easily can negate the best exhibits and programs. Conversely, a well-designed and maintained building can have a favorable impact on the visiting and supporting public, regardless of the quality of the science center's activities.

12

Security and Safety

In general, science centers do not have the priceless artifacts and works of art that art, history, and natural history museums must protect with large guard forces and elaborate security devices. However, they do have exhibits, collections, and a physical plant that must be shielded from theft, damage, and fire and people who must be protected from injury.

In 1974, the International Council of Museums (ICOM) established a Museum Security Committee because of the pressing need to stem museum losses, primarily in the art field. "The world's cultural legacy has suffered extraordinary losses through theft and misappropriation, vandalism, and environmental damage, fire, and flood," the committee pointed out in its security handbook, *Museum Security*.[1]

Security is the most important consideration in the administration of *any* museum, according to an ICOM manual dealing with the organization of museums;[2] yet, too many museums, including science and technology centers, fail to give appropriate attention to the security function.

"Museums, by their very nature, must be security conscious," explained G. Ellis Burcaw in an American Association for State and Local History planning guide. "Insurance is a poor substitute for preventive measures because the value of museum collections cannot really be expressed in financial terms. In guarding against theft, carelessness, and vandalism, the museum is placed in an awkward position because good public relations are so important to it. The museum wants to welcome all kinds of people in great numbers and to make them feel 'at home.' While being friendly and hospitable, the museum must also prevent any damage and disturbance the visitor might cause."[3]

Sometimes the security problems are as great from within as from without. Joseph M. Chapman, a private investigator specializing in art thefts, said: "I can't prove it, because most inside jobs are never reported, but my experience suggests that big museums probably have as many objects disappear from their storerooms as from exhibition areas."[4]

Unfortunately, many museums, including science centers, are lax about security and do not report or publicize their losses. Three Cezanne paintings valued at $3 million were stolen from a storage room at the Art Institute of Chicago in 1978. In addition to uncovering that many employees had keys to the supposedly secure room, the police investigation disclosed that in 1970 a $200,000 Chinese scroll was found missing but was never announced publicly for fear of encouraging further thefts.[5]

Security normally involves the protection of an institution's buildings, contents, staff, and visitors, but it also includes the care of the collections, insurance against financial loss, and physical protection against theft, vandalism, fire, and personal injury.

Security Force

The director of a science and technology center is ultimately responsible for the security of a museum, but the trustees must provide him or her "with sufficient financial resources to maintain protection, by means of technical equipment and a security staff or at least a chief security officer."[6]

In a small science center, the security authority usually is delegated to one of the staff members who has other responsibilities, as most small institutions simply cannot afford to have a full-time security officer. Medium-size and large science centers generally have one or more full-time security persons, including a security chief, who is responsible for planning, organizing, coordinating, and controlling all permanent and temporary security measures. According to the ICOM security handbook, "This means that he should be consulted on all matters that have any influence on security, such as museum remodeling plans, exhibitions, and the appointment of new staff. He, in turn, should advise the director on the advantages and disadvantages of various technical systems and on the latest improvements in security measures."[7]

The basis of any museum's security is physical guarding. Mechanical, electrical, and electronic devices can be extremely useful, but they cannot replace guards or security officers, especially with large numbers of visitors and curious employees. It generally is recommended that a science center hire and train its own security people instead of using an outside protection service. Such an approach gives the institution greater control over the loyalty, quality, training, and work assignments of its security force.

Traditionally, guards have come from the unskilled and/or retired ranks. But it is necessary to upgrade the quality, status, and pay of the security force if it is to be effective. A guard should be in good physical condition, trustworthy and dependable, and capable of dealing with the public and a somewhat monotonous routine.

Among the duties of a security officer are checking objects brought into the building, patrolling exhibit halls, setting alarm systems, controlling keys, inspecting doors and windows, directing traffic, maintaining order, investigating incidents, checking identification, hearing complaints, handling emergencies, and answering questions. Security procedures and assignments should be clearly spelled out for the security force and be respected by the staff and the public.

Security Systems

Security devices have become quite sophisticated and expensive, but it may not be necessary for the average science and technology center to make a major investment in mechanical, electrical, and electronic security equipment. If a science center does not have extensive or valuable collections, it may be sufficient to rely on a conscientious security force, effective security procedures, and a few security alarms.

In setting up a security system, Denis B. Alsford, author of the Canadian Museums Association security manual, suggests that museums should be aware that all security problems cannot be handled in the same manner and should seek advice from police and fire departments and possibly a security services firm.[8]

Jack Leo, a security consultant, believes the first step in establishing a good security system is "to be aware of the need." "Most of the time," he says, "this matter is never considered until after the museum has experienced a loss." It is also important to make "a survey of the weakness of the museum" and "to make the museum less vulnerable."[9]

A great deal can be done to improve security at relatively little expense to prevent or deter the casual thief. A wide variety of more expensive security equipment is available to thwart the professional burglar. An institution's needs, resources, and policies must dictate which approaches are most appropriate.

Basic Security Measures

Every science and technology center has certain basic security needs that can be met with a little planning, common sense, and a minimum budget. These security measures include such steps as good recordkeeping and inventory control; key and identification card control; checkroom, register, and package passes; secure screws, locks, doors, and windows; night lighting and low shrubbery; regular security checks; and a few alarms.

Recordkeeping and Inventory Control

A record of all objects in collections and exhibits and all equipment should be kept and use of such objects and equipment should be controlled. Whenever possible the collections should be fully documented with complete descriptions and photographs for inventory control and recovery purposes. It also is desirable to have a procedure and forms for the acquisition, loan, and disposal of collections and exhibits. Access to the artifacts and exhibit storage areas should be restricted, and periodic checks should be made to see that the materials are not missing, damaged, or threatened by environmental or other conditions.

Key and Identification Card Control

Many security difficulties at science and technology centers are related to the lack of a master key system and inadequate control of keys. The distribution of keys should be limited, and a careful record should be kept of key possession and use. Certain basic keys should be checked in and out daily for day or night use and kept in a locked cabinet for emergency purposes. When keys are lost, stolen, or misused, it usually is wise to change the locks as a precaution. In larger institutions, employee identification cards can be useful for the admission of staff members to restricted areas and to the building in off-hours. Identification tags or badges also may be used for temporary workers and authorized visitors to obtain access behind the scenes. In both instances it is vital to have full control over identification cards.

Checkroom, Register, and Package Passes

A checkroom at the entrance of a science center is a worthwhile investment for convenience and security reasons. Museums should have firm regulations on what can be brought into and taken out of the building. A guard stationed at the entrance should direct the public to check coats, umbrellas, containers, radios, and other objects that pose potential security hazards. A night register also should be established, and employees should be required to have package passes signed by their supervisors to remove parcels from the museum (and guards should be authorized to search such materials when appropriate).

Secure Screws, Locks, Doors, and Windows

A museum does not have to be a fortress to be secure, but it must have sound screws, locks, and doors. In exhibit cases, screws and locks should be used that cannot be sprung easily with a penknife, nail file, or screwdriver. Wherever possible, they also should be concealed. All doors and locks to secure areas should be of solid construction and be difficult to penetrate. Doors and frames should fit well; hinges should be on the inner side of the door and set screws should be used to hold the pins firmly in hinges; metal grills and shatterproof glass should be part of glass panel doors; and dead-bolt locks should be used rather than common spring-loaded, self-locking models that can be opened with plastic cards. When padlocks are used, they should be of high quality, with a one-piece body, saw-resistant shackle and hasp, and key lock rather than an automatic spring system. Windows also should fit well—both the sashes and the frames—and be of the metal casement type with locks if possible. Windows on the ground floor or basement level, particularly at the rear of the building, should be protected by steel bars or heavy wire mesh screening.

Night Lighting and Low Shrubbery

The use of outside and/or inside lighting can be an effective deterrent to crime. The installation of external light-sensitive controls and internal timing devices can make these almost automatic operations at relatively little cost. The outside lights and shrubbery should be located so that police patrols, neighbors, and passersby can see any intruders.

Regular Security Checks

All security systems should call for periodic patrols of the building both during the day and at night, with the museum's security force covering the inside and local police the outside. Security officers should check the exhibits and collections on display at opening and closing; maintain vigilance while the museum is open; examine secure storage areas periodically; make certain all visitors have left the building at closing; lock all doors and windows securely at night; and set all detection devices for overnight security. If an institution has one or more night watchmen, they should have specified security patrol routes, time clocks to punch, and clear instructions on who to admit, what can leave the building, and how to react in case of theft, fire, flooding, or other emergencies.

Alarms

A modest system of simple alarms may be appropriate for some science centers. They can be either a local alarm of bells, buzzers, sirens, or flashing lights or a remote silent alarm that alerts the local police, a central security station, or an institution's security patrol panel. Such alarms can be used to protect objects, rooms, and buildings, with the most common application being to guard valuable artifacts and exhibit objects, collection and exhibit storage rooms, offices and safes containing cash, and entrances, exits, and windows.

Security Devices

If circumstances require and the budget permits, a science and technology center may install a wide range of mechanical, electrical, and electronic security devices. These anti-intrusion detectors operate by indicating the presence or absence of or changes in certain physical phenomena, such as mechanical vibrations, electrical currents, sonic waves, optical and thermal rays, and electrical, magnetic, and electromagnetic fields. The devices vary greatly in sophistication, cost, and effectiveness.

Electrical and Magnetic Sensors

Built-in electrical wires, magnetic contact switches, and metal foil tapes are used frequently for perimeter doors and windows. Built-in wires can activate an alarm when the wire in a door or window is broken through forced entry. Magnetic contact switches make use of a magnetically actuated switch and a magnet to generate a signal when a door or window is opened. A metal foil tape glued around the edges of the window can set off an alarm when broken.

Vibration Detectors

Vibration detectors, such as piezoelectric devices and contact microphones, can be implanted in or affixed to windows, walls, ceilings, and floors to sense mechanical vibrations, such as drilling, chopping, or sawing.

Contact Mats

Contact mats, which also are known as pressure-sensitive mats or carpet detectors, set off an alarm as soon as someone steps on them. They normally are installed under carpeting, rugs, and doormats and are used primarily around sensitive objects in exhibit areas. When pressure is applied between two sets of conductors, an alarm is given.

Photoelectric Beams

Under this system, a field of infrared or laser beams is set up between a number of transmitters and receivers to form a protective screen. When an intruder interrupts any of the beams, a signal is generated.

Ultrasonic, Microwave, and Radar Motion Detectors

These space saturation systems are extremely sensitive to movement. An ultrasonic detector transmits and receives inaudible acoustic waves and uses the Doppler effect to distinguish between stationary and moving objects. The microwave system uses both electromagnetic waves and the Doppler effect. A radar method is based on a change in the intensity of the received radar wave caused by the absorption of energy of an object. Of the three, the microwave detector is considered less troublesome and more reliable.

Photographic Surveillance Systems

This category includes still and motion cameras and closed-circuit television. Still and motion pictures rarely are used in museums, but closed-circuit television surveillance can be extremely useful in exhibition halls, storage areas, and remote locations, such as rear exits and the receiving dock. Television units come in constant-scan and fixed-position models, with both regular- and low-light sensing cameras. They are most effective when combined with an audio signal generator that is activated when the camera detects movement.

Fixed-Point Sensors

In addition to using sealed display cases and secure fastenings, a number of sensors are available to protect objects on exhibit. These fixed-point or spot systems, which emit an alarm when an object is removed, include built-in wires; mechanical, contact, magnetic reed, and vibration switches; electromagnetic induction and ultrasonic devices; photoelectric beams; and displacement, weight, pressure, proximity, and audio sensors.

Fire Protection

Fire protection is closely related to security. In many instances it is the responsibility of the security chief. Other times, the operations director or building engineer is given the assignment.

Although most science and technology centers are not collection-oriented, fire still can endanger human lives and cause severe property damage. It is essential for every institution to develop a system for the prevention, detection, and extinguishing of fires. Such a plan should include an evacuation and fire-fighting procedure, regulations on smoking and the use of hazardous materials, fire and building code conformation, periodic fire inspections and drills, use of fire-resistant materials, correction of fire hazards, an alarm system, fire doors, a standpipe and hose system, fire and smoke detectors, an automatic water sprinkler system, portable fire extinguishers, and a training program on the use of extinguishers and other fire-fighting equipment.

Most museum fires begin behind the scenes rather than in the exhibition halls. They are the result of defective heating plants, unsafe handling of flammable liquids, faulty wiring, and careless smoking, frequently by employees. In developing a fire prevention plan, a science center should seek assistance from the fire department, insurance representatives, and suppliers and specialists in fire protection systems. Among the most important elements in any plan are fire alarms, detection systems, and extinguishing equipment.

Alarms

Every museum should have a fire alarm system, and every staff member should know how to sound the alarm or notify the fire department before attempting to extinguish a fire. An alarm signal connected to a central panel in the museum, a central station outside the museum, or directly to the fire department is much more effective than a locally sounding alarm. All manual alarm systems should be supplemented by automatic detection and alarm systems.

Fire Detection Systems

The two most common fire detection systems are thermal detectors and smoke detectors. Thermal detectors are activated by a sudden increase in a room's temperature. They generally are used in rooms with relatively low ceilings and set off an alarm when the ambient temperature rises above a certain point. Thermal devices can be a unit mounted on the ceiling or wiring or metal tubing usually at the junction of the ceiling and walls. Smoke detectors are a more recent development. They respond to the aerosols released during combustion and come in two forms: photoelectric and ionization detectors. The photoelectric type are more responsive to smoldering fires, and ionization units are more responsive to mostly invisible products generated by a flaming fire.

Fire Extinguishing Equipment

Fire extinguishers can be automatic systems, portabe units, or hose systems. Water sprinkler systems are the most common and effective automatic equipment. They are required in some cities. The typical sprinkler system has overhead pipes containing water under pressure, which is released through sprinklers activated by heat. The system also can serve as a thermal detection device by sounding an alarm when activated. Similar systems containing carbon dioxide, halogenated hydrocarbons, or dry chemicals are available for special types of fires. Portable fire extinguishers should be located throughout a museum, and employees should be trained in the use of the equipment. Extinguishers using water under high pressure or dry chemicals are preferred over soda-acid or carbon dioxide extinguishers. Hose systems are permanently connected to standpipes and should be located throughout a science center. The most common types of fires in museums involve ordinary combustibles such as paper, textiles, and wood (extinguished by cooling, blanketing, or wetting); oils, greases, paints, and flammable liquids (extinguished by smothering or blanketing); energized electrical equipment (requires a nonconducting extinguishing agent).[10] These fires and the extinguishers for putting them out are rated as Class A, B, and C, respectively.

Safety Precautions

Every museum is responsible for the safety of its employees, visitors, and others who utilize its facilities. The chief security officer or the building engineer is usually assigned to be responsible for safety, but a science center should also appoint a safety committee and institute an accident prevention program.

Safety problems can be caused by overcrowding, running, falling, or bumping, sharp obstacles, inadequate lighting, fights, slippery floors, working with hazardous equipment or materials, steep steps, carelessness, vandalism, moving exhibits, loose carpets, poor sanitation, icy sidewalks, fires, chemical fumes, pickpocketing, faulty furniture, or misplaced tools. Some have security implications; others require better supervision or procedures. All have a bearing on personal safety and the institution's liability insurance.

A science center should survey the safety hazards of its building and take steps to correct problems that might cause personal harm.

This may involve adding antislip strips to steps, improving the lighting in dark areas, placing protective covers over dangerous machinery, using warning signs when mopping floors or making repairs in public areas, installing railings on steep stairs, and fastening loose carpeting. An attempt should be made to provide benches and other seating for the elderly and fatigued. Federal regulations also require that museums be made more accessible to the handicapped, calling for ramps, special restroom stalls, and even elevators.

All institutions should have a first aid room and several persons trained to handle minor accidents, and arrangements should be made to rush serious cases to the hospital. A system of written accident reports and followups also should be incorporated into the museum's operating procedures.

13

Public Relations and Publications

Public Relations Framework

Every science and technology center depends on publicity, publications, and promotion to attract and serve visitors, members, contributors, trustees, employees, volunteers, teachers, and taxpayers. But an institution's communications program should extend beyond publicity to include publications, direct mail, a speaker's bureau, films, community relations, audience development, special events, and other activities designed to enhance the museum's position in the community, region, and/or nation.

The activities designed to improve a museum's image or acceptance generally are the responsibility of a public relations, public information, or public affairs department. In most instances, the department consists of a director and a secretary, but in larger institutions the public relations staff may include one or more writers, a publications editor, a photographer, and/or a graphic designer. Sometimes the public relations function is part of the development office. At still other institutions, a public relations committee is comprised of members from the governing board, membership organization, and/or volunteers' group. In a few cases the public relations function is carried out entirely by volunteers.

At all times public relations should receive the full attention and support of the museum director. He or she should be involved in approving major news releases, publications, and other PR activities and should rely on the public relations director for guidance and assistance in decisions, announcements, speeches, and other actions affecting public attitudes toward the institution.

There are two sides to public relations: the everyday visitor services and the formal public relations program. "All staff members who meet visitors are involved in personal public relations. It is the impression these staff members make on the visitor that will bring forth favorable or unfavorable comments, either to the staff or outside the museum to the general public."[1] The secret to personal public relations is "good manners and a smile," according to the Canadian museum management manual. Visitors should be made to "feel at home," and every effort should be directed to give the museum a "receptive atmosphere" for learning.[2]

Formal public relations is an organized process dealing with media relations, publications preparation, and special promotional activities. The importance of this function is to make "the public aware of the programs and facilities of the museum, and of what the museum requires in order to continue to grow. A successful museum plans a systematic program of communications which informs those connected with the museum of important news, aids in interpreting exhibits, publicizes programs, and generates excitement in the community concerning museum activities."[3]

The size, budget, and activities of public relations departments vary greatly, but nearly all are concerned primarily with publicity and publications.

Public Relations Techniques

The public relations techniques utilized by a science center usually fall into three broad categories: media publicity, museum publications, and special promotional activities. The degree and nature of usage fluctuates greatly from one institution to another.

Media Publicity

Publicity usually refers to media coverage, in newspapers and magazines and over radio and television. The goal of every public relations staff is to obtain as much favorable publicity as possible. However, because of the limited space and time available and the great competition by virtually every organization in the community, only a fraction of any public relations department's efforts results in being published or aired.

In general, the more professional the public relations work, the better the chances of success. But the principal determinant in publicity usually is the newsworthiness or human interest of the story or program. Not all publicity is planned or favorable, but every attempt should be made to inform and respond to the press. Key elements in media relations are accuracy, reliability, and integrity. A public relations staff that develops a reputation for withholding information or covering up on thefts, fires, accidents, and other unpleasant subjects can lose its effectiveness.

News Releases

News releases are the principal means of communicating information to newspapers, magazines, news services, and radio and television stations. Also called press releases and "handouts," they are written in a style similar to newspaper articles, with the most important information first. News releases usually are one to three typewritten, double-spaced pages in length and are mailed first class, delivered personally to news offices or distributed at news conferences and previews. At science centers, news releases generally are concerned with new exhibits, special events, educational programs, expansion plans, staff appointments, unusual artifacts, and research findings.

Feature Stories

Instead of sending the same information to all media, a public relations staff member sometimes will work with a single newswriter on a feature story that is not considered "hard" news. The idea for such a story may originate with the PR person or media and generally deals with people having unusual jobs, behind-the-scenes preparation for a new exhibit or event, background information on a new or old collection, operation of sophisticated instrumentation, experience on a field trip, and other so-called "soft" news. A public relations staff member should not show favoritism to a particular medium, but cooperating on "exclusives" of this type is acceptable if evenly distributed among the media over a period of time. In some cases feature stories are written by PR personnel and edited by the media, but more often such articles are researched and written by a media staffer with assistance from the center's public relations department.

News Photographs

Many science centers employ a photographer or staff member to take pictures for use in news releases and museum publications. Others hire a professional or free-lance photographer when needed. News photographs generally are eight-by-ten-inch glossy black-and-white photos, although they sometimes are made to column width for publications produced by the offset process. Because of cost, distribution of photographs usually is selective. However, it is possible to obtain hundreds of copies at relatively low cost from certain national photo processing firms if there is enough lead time. Color transparencies frequently are taken for Sunday supplements, magazines, museum publications, and slide presentations. All photographs and transparencies should be mailed with cardboard support.

News Conferences and Previews

A news or press conference sometimes is held as a convenient way to present a significant news event or introduce an important person to all media at the same time. A slightly different version of the news conference is the press or exhibit preview, frequently used to permit media to view a special exhibition before it is opened to the public. In both instances, a press kit containing a news release, background information, and one or more photographs often is prepared by the public relations staff for attendees. A science center must be careful not to call unnecessary press conferences that might discourage media interest on important occasions in the future.

Radio and Television

Radio and television provide five types of publicity opportunities: spot news, interview programs, filler programming, public service announcements, and programs presented by the museum. Spot news is handled on a station's news program and relies heavily on news releases, feature stories, conferences, and exhibit previews. Most stations also have regularly scheduled talk or interview shows on which the public relations staff can place guests from the science center. It also may be possible to have a film or taped program on the air when a station has to fill time because of a delay, postponement, or cancellation of a program. In the United States and some other countries, government regulations require that radio and TV stations devote a certain amount of their time to public service programming. This usually takes the form of brief public service announcements and occasionally a series of programs presented by community service organizations.

Museum Publications

Publications should be an important part of any science center's public relations program. They supplement media publicity efforts by enabling an institution to tell its story in greater detail and to maintain regular contact with visitors, members, contributors, trustees, employees, and volunteers. A center may provide information leaflets, guidebooks and catalogs, calendars of events, newsletters and magazines, annual reports, and various special-purpose publications for educational, membership, and fund-raising activities.

Some public relations offices have a publication editor and/or designer, but most museum publications are written by the PR director or staff and are designed with the help of the exhibits staff, a free-lance artist, or a printer. Every effort should be made to have readable, attractive, and effective publications because they have considerable influence on the attendance at and the support and impact of the institution.

Informational Leaflets

Every science center should have one or more informational leaflets that concisely describe the institution, its offerings, and its location. These publications, frequently called "giveaways," are intended for free and widespread distribution at the museum entrance, through the mails, and at tourist information counters, schools, hotels and motels, and travel agencies. These leaflets range from inexpensive three-panel folders to elaborate four-color pieces and normally are designed to fit in a standard #10 envelope.

Guidebooks and Catalogs

Science centers also publish various types of guidebooks and catalogs, most of which are intended to be sold in museum stores. A guidebook may cover all or part of a science center's exhibits, frequently with a floor plan and numerous photographs. On sponsored or temporary exhibits, a descriptive folder or guide may be distributed free to visitors. In many cases, because of the maintenance problem, such literature is available only at the information counter. An expanded version of the guidebook is the catalog, which may deal with an institution's total collection or an individual exhibition. A catalog frequently is prepared by a curator rather than the public relations staff and contains information and illustrations on specific objects on display.

Calendar of Events

A periodic calendar of events distributed at the museum and/or to those on a mailing list can be useful in building attendance for new exhibits, educational programs, and films. Such publications usually are issued monthly or quarterly and are designed to be self-mailers and possibly to be placed on bulletin boards. Another form of calendar is the annual calendar for use throughout the year. Such calendars frequently contain large black-and-white or color photographs pertaining to the museum, dates for important museum events, and space for personal notations. Annual calendars generally are mailed free to members, contributors, and schools and are often sold to others.

Newsletters and Magazines

Nearly every science center has some form of periodic newsletter, magazine, or journal that goes to members, contributors, educators, employees, and/or other groups of interest to the institution. Newsletters usually are four- or eight-page, 8½-by-11-inch mininewspapers with news stories and photographs of activities and events at the institution. They normally are published monthly, bimonthly, or quarterly. A few science centers have their own magazine or journal—generally published quarterly—that ranges from twelve to thirty-two pages and contains longer, feature-oriented articles on museum activities and/or the museum field. Science centers with curators that conduct research may report on their investigations and discoveries in such publications or in occasional research bulletins.

Annual Reports and Fact Books

An increasing number of museums are publishing annual reports, either separately or as part of newsletters and magazines, to report on their progress, account for expenditures, recognize those who have contributed or assisted, and point out the direction and needs of the institution. Museums that do not have an annual report sometimes prepare an annual "fact book" for media and other uses on the history, philosophy, operation, funding, exhibits, and collections of the museum.

Special-Purpose Literature

A public relations department may be involved in preparing a wide range of other literature for educational, membership, fund-raising, and other purposes. These include teachers' guides for group visitations, pledge cards, brochures to promote capital campaigns, and even booklets and books for anniversaries and other special occasions.

Special Promotional Activities

Most public relations staffs are involved in many activities beyond media publicity and publications. These efforts may include such promotional activities as signs and posters, films and slide presentations, a speakers' bureau, special tours and events, audience development, and community relations. Again, the needs and funds must determine what is appropriate and possible.

Signs and Posters

Signage inside and outside the building usually is the responsibility of the public relations department, although the design and/or production may be handled by other offices or contractors. A uniform system of directional and identification signs that are attractive, readable, and meaningful is a necessity for every institution. It also is helpful to have signs identifying the building and directing traffic to the institution from main streets and highways. Another possibility is a billboard sign in a prominent location that can be rented at nominal cost as a public service. Some science centers also distribute for free or sell institutional and/or exhibition posters to promote a temporary or traveling exhibition as well as to serve as mementos of the exhibition.

Films and Slide Presentations

Although expensive, a motion picture on preparing for a field trip, viewing particular exhibits and collections, and visiting the science center can be used in schools, on television, and at community and fund-raising meetings. Films should be 16 mm to keep costs down and avoid difficulties with the projectionists' union. Slide shows—usually 35 mm—can be used for similar purposes. They cost less, have greater flexibility, and do not become outdated as rapidly as motion pictures. Three extensions of these audiovisual techniques are film strips, videotapes, and multimedia presentations. Film strips sometimes are useful in schools, talks, or special presentations, while videotapes generally are aimed at classroom or television use. Multimedia presentations usually consist of three or more projectors and screens showing films and/or slides for orientation, fund-raising, or other purposes.

Speakers' Bureau

Museums frequently receive requests for speakers, and such inquiries usually are handled through the public relations department. Sometimes the PR office makes a conscious effort to place speakers and operates a "speakers' bureau." Such efforts may include publishing a list of speakers with possible topics or mailing a letter and/or literature to schools, service and women's clubs, professional and technical societies, church groups, and fraternal organizations. In some instances, films and slide presentations are used in this program.

Special Tours and Events

Although routine tours of exhibits and the building are handled by the education or program department, certain VIP, media, and special tours are handled by public relations personnel. In addition, the public relations department may be responsible for organizing some or all special events, such as exhibition openings, news conferences, evening open houses, film showings, and kickoff luncheons and dinners.

Audience Development

The words "audience development" are relatively new in the museum field, but at least one science center—the Franklin Institute Science Museum in Philadelphia—calls its public relations officer the director of audience development. Use of the phrase reflects the increasing concern of museums with increasing attendance, membership, contributions, and other support. The Franklin Institute engages in a number of special promotions, such as scientific demonstrations at baseball games, mini-exhibits in shopping malls, tie-ins with commercial products, and science fairs on parkways, to attract attention and support. Other science centers, such as the Museum of Science and Industry in Chicago, present special holiday and minority programs to broaden the museum's appeal and service.

Community Relations

Many of the public relations activities mentioned earlier are utilized in museum efforts to improve community relations. In addition to publicity, publications, and general promotion, some science centers have advisory committees, neighborhood events, ethnic programs, activities for senior citizens and the handicapped, outreach programs, science fairs, and other activities to cultivate community interest and improve rapport.

Visitors to the Ontario Science Centre in
Toronto learn about computers through
interaction.

Learning about science first hand is the focus of most science centers. Here a youngster finds out about electricity at the Oregon Museum of Science and Industry in Portland.

The first technical museum was the Musée National des Techniques du Conservatoire National des Arts et Métiers (National Technical Museum of the National Conservatory of Arts and Trades), founded in Paris in 1794. It opened in an eleventh-century Benedictine priory when the French Revolution ended in 1799.

The Science Museum in London was a spinoff from the Crystal Palace Exhibition in 1851. Its collections originally were part of the South Kensington Museum of Industrial Arts (later renamed the Victoria and Albert Museum). In 1909, the scientific and technological objects were separated from the decorative art collections to form the science museum. This photograph shows the operating steam engines in the entrance hall of the museum.

The Deutsches Museum in Munich was the first contemporary science and technology museum. Founded in 1903, it now is located on an island in the heart of the German city.

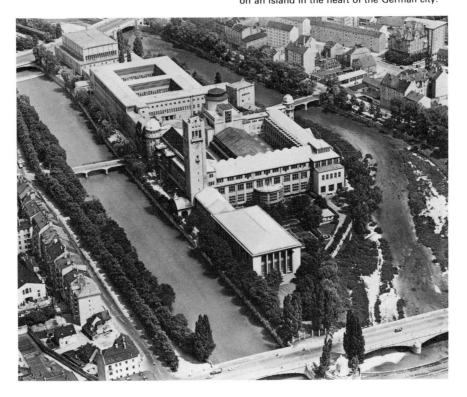

The first successful contemporary science and technology center in the United States was the Museum of Science and Industry in Chicago, housed in the Palace of Fine Arts. Founded by Julius Rosenwald in 1926, the museum opened to the public in 1933. It remains one of the most popular museums in the world, with an annual attendance of about four million.

In the late 1960s, the University of California, Berkeley, opened the Lawrence Hall of Science, with an emphasis on science education research and development as well as the use of computers.

The Ontario Science Centre in Toronto was opened in 1969 by the Province of Ontario to mark the centennial of Canada's Confederation. It is located in a split-level complex of three interconnected buildings.

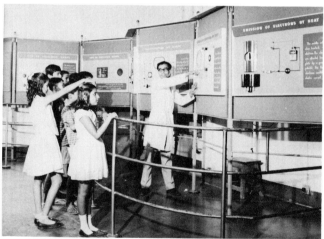

Field trips are a part of many educational programs at science centers. This photo was taken during a geological enrichment outing organized by the Oregon Museum of Science and Industry in Portland.

Basic science education is one of the objectives of India's science and technology centers. In this photograph, the fundamentals of electronics are explained to students at the Birla Industrial and Technological Museum in Calcutta.

Most science centers present scientific demonstrations and need trained personnel to present the programs, such as this one at the Alfa Cultural Center in Monterrey, Mexico.

Educational tours, such as this one of the observatory at the Buhl Planetarium and Institute of Popular Science in Pittsburgh, are conducted by staff members and volunteers.

World's fair buildings have been inherited by
a number of science centers. The U.S. Science
Pavilion from the 1962 Seattle World's Fair,
for example, was converted into the Pacific
Science Center.

A new addition to the Museum of Science in Boston has three levels and makes use of large open spaces and escalators. The floor plan gives great flexibility, and the escalators make the exhibits easily accessible.

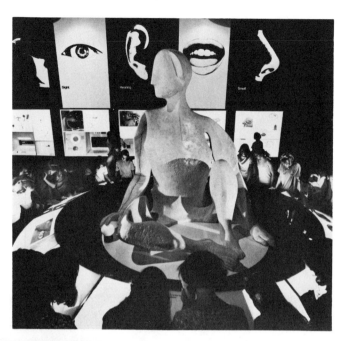

Design has become as important as content in many science exhibits. This exhibit explains the senses at the California Museum of Science and Industry in Los Angeles and was designed to attract as well as inform the visiting public.

What is the purpose of the exhibit and how can the information best be transmitted? Such questions must be considered in designing any exhibit. At the Palais de la Decouverte in Paris participatory pulleys demonstrate the principles of simple machines.

Large-screen "Imax" and "Omnimax" theaters are operated by a number of science centers. The Reuben H. Fleet Space Theater and Science Center in San Diego has a hemispheric screen that is seventy-six feet in diameter and is inclined twenty-five degrees from the horizontal. The screen creates a "you-are-there" effect for the high-fidelity motion pictures.

Traveling science shows are presented in schools by many science centers, sometimes for a fee. Here a chemistry program is being given in a Philadelphia area elementary school by a Franklin Institute Science Museum demonstrator.

The Pacific Science Center in Seattle presents an annual "Science Circus" during the Christmas holiday season. Up to 50,000 people have attended the special program, which includes "Ralph the Robot," science "magic," participatory exhibits, make-it-yourself experiences, and science demonstrations, as shown in the illustration.

As part of the Franklin Institute Science Museum's outreach program, the museum operates the "Museum on the Mall" at a downtown shopping center, giving the public a sampling of the science exhibits and demonstrations at the museum.

III

Exhibits and Programs

14

Collections

Objects Versus Exhibits

The collection of objects of intrinsic value traditionally has been the prime function of museums. In fact, a "museum" has been defined as "an institution devoted to the procurement, care, and display of objects of lasting interest or value."[1]

In a 1957 museum management guide written by Carl F. Guthe for the American Association of Museums, he emphasized:

The heart of the museum is in its collections. An organization may be an art or history center, a community cultural activity, or a children's recreational service, but it cannot be a museum without permanent collections, and the manner in which they are cared for and used by the museum, determines its standing among other museums, and its prestige in its community.

The organization, the management, and the activities of the museum exist solely to insure the continuous adequate care of the materials in the collections and their effective use for cultural and educational purposes. The first obligation of a museum is to recognize and assume the responsibilities inherent in the possession of its collections, which are held in trust for the benefit of the present and future citizens of the community.[2]

In 1969, George MacBeath and S. James Gooding stated in the Canadian Museums Association booklet *Basic Museum Management* that "We have seen that the museum is a unique institution in that it performs its functions by the collection and use of artifacts or specimens, or both. Without collections there can be no museums. Collecting and collections must be at the heart of a museum's operations, and research the keystone to the authenticating and interpreting of these collections."[3]

But practices, definitions, and philosophies change over the years. As science and technology centers and other types of museums placed less emphasis on "collections" and more on "exhibits," the museum profession began to recognize the need for revising its traditional outlook on what is or is not a museum. This evolution was best illustrated in 1975 when the American Association of Museums changed its eligibility ground rules to enable science and technology centers, planetariums, art centers, and other institutions that do not necessarily both own and utilize objects of intrinsic value to become accredited.

Nearly all museums, including science centers, still collect and have collections of objects. The principal difference is the role assigned to artifacts and specimens in the operation of the institutions. At many science and technology centers, collections have been replaced by exhibits specifically developed for educational purposes. Thus, objects have assumed a secondary or supporting role. This transposition of importance has had profound implications for the organizing, staffing, and programming of science centers.

As a result, many science centers have relatively few artifacts in their collections and operate without curatorial staffs. They also conduct little or no research and do not publish learned papers. Their exhibits contain few original objects and cases with hands-off signs. They do not approach subject areas sequentially or insist on the usual museum decorum.

Science and technology centers have broadened the concept of museums and have sought to make museum-going both enlightening and entertaining. They have changed the rules of the game, and other types of museums are adapting their philosophies and techniques.

It has not been an easy struggle for science centers or other museums that have attempted to make the transition, as evidenced by the criticism heaped on the Ontario Science Centre in Toronto at its founding and the internal squabble at the staid British Museum of Natural History in London.

Many Canadian professionals criticized the decision to make the Ontario Science Centre a contemporary science and technology center instead of a traditional science and technology museum. One such castigation was an article by Duncan F. Cameron, president of Janus Museum Consultants Ltd., and later director of the Brooklyn Museum, who stated that "it contains a veritable chaos of science exhibits mixed with industrial and technological exhibits sponsored by corporations."[4]

In 1975, the British Museum trustees created a "Department of Public Services" to improve the diffusion of knowledge to the general public through more interesting and effective exhibits. Dr. Roger S. Miles, head of the new department, wanted "to make the museum an exciting place where the layman can enjoy exploring and discovering natural history. We need to attract an audience that is representative of the general public. We must recognize the need to motivate visitors."[5]

The new publicly oriented exhibit policy has met with strong opposition from the museum's curatorial staff and the academic community. The first exhibition—dealing with human biology—caused a turmoil in British museum circles. Professor B. Halstead, a geologist at the University of Reading, wrote that the new exhibition approach clearly poses "a fundamental dichotomy that exists in the concept of the role of the museum." He severely criticized the Department of Public Services' position that "its duty is primarily to serve the general public, and more than that, not to provide mere displays of the materials housed, but to communicate ideas and concepts."[6]

Perhaps Professor Halstead has put his finger on one of the major drawbacks of collection-oriented museums: many do not communicate ideas and concepts or they are ineffective in doing so. This may account for the great popularity of science and technology centers, which are more concerned with communicating than collecting.

Although a few science centers exist without collections in the usual sense, virtually all make some use of objects in their exhibits, education programs, and other activities. They have found that it is possible to bridge the so-called "dichotomy" between collections and exhibits.

A Collecting Policy

A museum collects objects for two reasons, according to Carl E. Guthe: "First, it is a repository for objects which must be preserved because of their aesthetic, historic, or scientific importance . . . . Secondly, the museum is a storehouse for materials which have educational usefulness."[7] Many objects belong in both categories, but not all objects should be added to any museum's collections. A museum does not have the space, need, or budget to collect everything. Therefore, it is necessary to make choices or, in other words, to have a collecting policy.

Many museums, including science centers, begin with the offer or acquisition of a large private collection. Since the character and purpose of an institution frequently are determined to a large degree by early collection acquisitions, it is essential that a collecting policy be determined early in the life of a museum. It was, for example, the natural history collections of a wealthy London physician, Sir Hans Sloane, that were bequeathed to the British nation in 1753 that formed the basis of the troubled British Museum. The Henry Ford Museum and Greenfield Village in Dearborn are the result of the auto industrialist's collecting interests and support.

More often, science and technology centers are not faced with such major choices. Rather, it is an accumulation of smaller collections—even individual objects—that affect the nature and volume of science museum collections. Obviously, certain types of science centers, such as those that specialize in natural history, will have large systematic collections. Others may not need more than a small collection of objects.

Most people are collectors, and they often call museums when cleaning out their attics, basements, or garages or disposing of their collections. They will offer to give or sell to a museum an old car, vacuum cleaner, sewing machine, typewriter, or furniture or contribute old photographs, clothing, newspaper, or butterfly collections. A science center must have guidelines for responding to these and other collecting opportunities. Most important, the museum must be sure that all acquisitions are legal. The professional staff also must know what kinds of objects should be sought to supplement or fill voids in collections. It costs money to store, catalog, care for, and use artifacts. Collections require space, personnel, equipment, and materials that make selectivity a necessity.

A science and technology center also should state procedures for handling gifts, purchases, loans, and disposal of artifacts in its collecting policy. In accepting a gift, a museum should not agree to keep the gift items together, to exhibit them at all times, to associate the donor's name with objects every time they are used, or to keep the objects forever. It is better to lose a potential contribution than to commit the institution to such impractical stipulations.

A museum must have the flexibility to utilize its collections in the most efficient and effective manner and not be committed to the perpetual care or use of anyone's donated objects. To avoid any misunderstanding, a standard gift acceptance form or letter should be developed that states that the gift is for unrestricted purposes and does not come with any stipulations on the use or disposition of the objects.

Although most science centers buy few objects in assembling collections, they must be certain that the objects are genuine, were not stolen or smuggled into the country illegally, are fully documented, and are reasonably priced. They should require a bill of sale and documentation papers. If a museum is not qualified to determine the authenticity or value of an object or collection, it should make use of a competent and disinterested appraiser.

A museum should not appraise the value of gifts for the benefit of donors. Most museums require donors to obtain such appraisals from independent and knowledgeable sources. Museums also are contacted by the public to appraise objects they have found or uncovered in the attic. The best policy is to say the institution does not give appraisals.

Sometimes, a science center will have an opportunity to obtain an object, collection, or exhibit on loan. In general, it should avoid long-term commitments, and any agreements should be reviewed periodically. Most museums refuse the types of loans that are designed primarily to provide exposure or free storage for an owner's questionable collection or exhibit. On the other hand, some excellent objects can be obtained on loan for exhibitions organized by the staff, and many worthwhile traveling exhibits are available for temporary showings. Some museum exhibits funded by companies and government agencies also are technically on loan, since they are still owned by the sponsors, but museums usually have control over the content, maintenance, and display of the exhibits. Again, it is important to have a written agreement clearly spelling out various responsibilities for loans. The same applies to loans made by the museum to others, whether they are museums, shopping malls, schools, companies, or individuals.

Some museums never dispose of anything from their collections, even though they probably should do some housecleaning from time to time. The disposal of surplus materials requires exceptional care. A museum cannot afford to be accused of selling off its "valuable" collections. Therefore, most museums have a policy requiring study and justification for deaccessing objects from collections as well as approval of the museum director and/or governing board. Among the objects usually discarded, traded, or sold are duplicated, worthless, or artifacts or specimens that do not fit within the defined scope and purpose of the institution.

A collecting policy should be comprehensive and include as many of the foregoing points as possible. It should provide guidance with some flexibility and lay the foundation for institutional procedures on the recordkeeping, care, and use of the collections.

Collections Recordkeeping

Systematic recordkeeping is an essential part of any museum with artifacts and specimens. Documentation for an object is almost as important as the object itself. Without reliable identification and other recordkeeping, a museum's collections are of little use.

Collection recordkeeping takes place in the registrar's office, which is responsible for the "recording of all objects that enter and leave the building. From this follows the responsibility for their safe handling and storage while they are being recorded, for their unpacking, packing, and transportation, and the necessity for keeping track of their movement at all times."[8]

The purpose of a records system is to keep track of every object in the museum's possession; to gather and maintain all documents pertaining to the object's authenticity, description, origin, importance, value, and use; and to locate an object whenever needed. Accessioning, registration, and cataloging are three types of records kept by most science and technology centers. Accessioning and registration usually are carried out manually by the museum registrar, and the methods are similar regardless of the type or size of institution. Cataloging can differ dramatically from museum to museum.

Accessioning

The accession record or register is the key to any systematic recordkeeping system. It is here that all items in a museum's collections are entered first. Each accession receives a number, whether the collection consists of one or more objects.

The accession record may be kept in a bound or loose-leaf volume and/or in a card file. Although sometimes handwritten, a typed record is preferable for legibility. When typed on sheets, it also is possible to make a duplicate copy that can be bound and kept in a vault. Some museums prefer to place all accession records on cards for greater flexibility. In such cases, the cards should be kept in a locked file.

An accession file records the following data: accession number, name and description of object, origin, source, and nature of acquisition, date of receipt, and location in the museum (or away from the museum if on loan).

The accession numbering system is sequential, but there are variations. In most cases, numbers consisting of several parts are used. The first part gives the year in which an object or a collection is accessioned. For example, *79* would stand for 1979. A second group of numbers following a decimal point, such as *79.25,* would indicate that the object or collection is the twenty-fifth accession of the year. If the accession is composed of more than one object, a third number would be added after another decimal point for each object, such as *79.25.1, 79.25.2,* etc. Sometimes an object has two or more parts, such as snowshoes, and it is necessary to add a letter to the number to indicate several parts, such as *79.25.2a, 79.25.2b.*

Some museums prefer to operate with one number and use the accessioning number for registration, cataloging, and long-term loans. With loans, the letter *L* is added before the number; for example, *L79.25.2a.*

The matter of accepting and rejecting objects for a museum's collections normally is decided by a curator, an accessions committee, and/or the museum director. Rejections of donations can become delicate, but the task is much easier with a collecting policy and a clear understanding of what the museum needs and wants in the way of artifacts and specimens.

Registration

The purpose of registration is to provide a fast, concise, and permanent means of identifying each object in a museum's collections. It differs from cataloging, which seeks to classify objects methodically and with considerable descriptive detail. The accessioning and registration functions always are the responsibility of the registrar, while the handling of collection cataloging may be shared with the curatorial staff.

A registrar's job can become complex. When a gift or loan is made to the museum, the registrar must accurately identify the object and mail a temporary receipt to the donor. If a curator or the museum's accession department rejects the object, the registrar must mail a letter of regret and the object back to the donor. If accepted, the registrar enters the object into the accession register and assigns it an accession number, which is placed on the object and is used in all registration files. The registrar then completes a gift or loan form, important legal documents are signed by an authorized museum official, and copies of both are mailed to the donor.[9]

Other functions usually assigned to the registrar are managing collections in storage, shipping and receiving artifacts and exhibits, and handling insurance and customs matters relating to collections and exhibits.

The registration record should contain as much information as possible about the history of an object and how it was obtained. The registrar may question donors or curators or refer to newspaper articles, books, catalogs, sales slips, or transmittal papers for such information.

The registration system may have only accession, document, and source files or may include location, catalog, and classification files. In small museums or institutions with few collections, only the first three may be necessary. All of the files, except the document file, can be placed on cards and kept in file cabinets.

The accession record was described earlier. The document file, as the name implies, contains all the supporting material on each accession; such materials usually are kept in manila envelopes or file folders and are filed in numerical order. The cards of the source file, which contain the names of donors, lenders, or vendors as well as the registration number and perhaps a brief description of each accession, are stored alphabetically by source. A location file records the registration number and title of each object. It is arranged in numerical order and is intended to serve as a quick reference source. The classification and catalog files are primarily for the use of curators and researchers. The catalog file is the entire collection documentation arranged by a system other than accession numbers, while the classification file is arranged by category headings with cross references.

Cataloging

A museum's catalog card file is similar to a library card system. It contains basic information about an object, such as its registration number, title, description, history, origin, designer or artist, dimensions, type, classification, and information on the techniques and materials used to construction the object and on how it was acquired. The first card frequently is followed by a card with a photograph of the object and other cards with supplemental information.

Such cataloging requires considerable time and effort and the close cooperation of the curatorial staff and the registrar's office. It also has some inherent problems, such as deciding on the best classification system to meet multiple uses and maintaining a reasonable consistency in the classificatory terms place at the top of the cards for filing purposes.

Catalog cards also present two major problems for users. "The labor involved in preparing, filing, and controlling card files with more than three or four cards per subject generally cannot be justified in terms of the use made of the added files," and "because it is usually cheaper to duplicate a single master card than to retype additional cards for specific purposes, all the cards usually carry all the information on an object," thereby producing data redundancy.[10]

It is for these reasons—and the greater storage capacity and faster retrieval—that an increasing number of museums are shifting to computer systems for cataloging and other registration records. Because of the high cost of conversion to computers, some museums also use the computer system for payroll processing, budgetary control, membership lists, development lists, library catalogs, inventory control, attendance figures, and publication and school mailing lists.

Care of Collections

Professional storage, conservation, and handling of collections should be an essential part of any museum with artifacts and specimens. It does not make sense to subject collections to deterioration and damage after going to such great effort, and possibly expense, to obtain objects for study and display. Yet the caring of collections continues to be a problem for science and technology centers. Insufficient attention, manpower, and funds are devoted to this important aspect of collections at many institutions.

Storage

Every science center with collections should have one or more storerooms for the safekeeping of artifacts and specimens. The rooms should have controlled access under the supervision of the registrar, and nothing should be added or removed without his or her approval.

Storage rooms generally should have fire and security alarm systems, temperature and humidity controls, storage shelves and cabinets, adequate work space and lighting, and be kept clean and uncluttered. They also should be fumigated periodically. Only staff members concerned with the collections should be permitted in storerooms, and, if possible, only personnel trained in handling museum objects should move objects. The collections should be arranged systematically, with the registration number viewable, so that they can be found easily.

Many types of shelving and cabinets are available commercially to museums—standard steel and wooden shelves, racked sliding trays, and sliding screen panels suspended from the ceiling or supported by the floor—or wooden shelves can be built by the museum carpenter.

Objects should be stored properly. For example, insect-prone textiles and natural history objects should be kept in reasonably airtight cases; fragile articles should be braced or cushioned in their containers to prevent accidental breakage; large objects should be covered with muslin or plastic sheets; and small objects should be placed in paper, cloth, or plastic bags or small boxes to keep them together.

G. Ellis Burcaw points out in his museum manual that the seven most common causes of collection damage are human loss (theft, vandalism, careless usage, accidents); environmental damage; climatic changes; radiant energy; biological damage; faulty procedures; and disasters.[11] Since all materials deteriorate, the best possible conditions for the longest life of each object must be maintained. Museums have an obligation to donors and society for the permanent preservation of its collections, and it cannot afford to have objects damaged or ruined by carelessness or neglect. Unlike most possessions, objects in museum collections frequently cannot be replaced.

Conservation

A museum's conservation responsibility includes all the procedures for keeping objects in good physical condition. It begins with proper storage and periodic inspections but may require corrective action when deterioration sets in. Unfortunately, few science and technology centers, except for the larger natural history museums, have conservators, restorers, and other such specialists on their staffs, but every collection-oriented institution must be vigilant, take precautionary steps, and call on outside experts when needed. It also may be possible to train a staff member to assist with certain types of conservation problems.

A science center must be aware that various materials require different types of treatment to preserve them from insects, light, temperature, humidity changes, and other deteriorating influences. Paintings should be cleaned periodically but by someone who is experienced in the field, and documents should be kept flat, unfolded, and away from light to avoid cracking and discoloration. Transparent cellophane tape, wire clothes hangers, and flimsy paper boxes should be avoided; cellophane tape shrinks and dries out, wire hangers rust and stain, and inadequate boxes collapse.

A conservator is a craftsman or artist whose purpose is to slow down destructive forces and possibly restore an object to something approaching its original appearance. A restorer's work ranges from replacing or repairing broken parts to giving the appearance of a new piece. Both are professionals more often found in art and history museums than in science centers.

Handling

More damage probably occurs from improper handling of collections than all other sources combined. Human error resulting from carelessness, ignorance, or arrogance has left its toll on collections of all types.

Science centers should have strict regulations regarding the handling of objects, from the moment of their arrival to their placement in storage, use in an exhibit, or packaging for shipment. Whenever possible, only personnel with training in the handling of artifacts and specimens should be permitted to move, uncrate, or package collections.

"The greatest care must be taken when moving an object within your own building," cautions one museum guide. "Use two hands and support the object from the bottom. Never assume that the handle of a cup or pitcher is solid; never assume that the wires on a picture are secure; never assume that part of an object is able to support the whole object, for such carelessness is the cause of most damage."[12]

The moving of crated collections in and out of a museum can be troublesome, particularly if the objects are not packed properly. Poor packing also can result in difficulties during shipment. Whenever possible, custom-made crates should be used for shipping both objects and exhibits.

Use of Collections

Collections in museums are used for research, exhibits, teaching, and loans. Each function requires artifacts and specimens that are authentic, adequately documented, and in good condition. At science centers, the objects are used primarily for exhibits and teaching purposes.

Research

One of the prime functions of traditional museums is the conduct of research, usually based on objects in museum collections. These original investigations are rare at most science and technology centers, except for those that are collection-oriented, such as natural history museums.

Typically the research is performed by curators and reported in scientific journals. Information uncovered in such studies frequently is incorporated in exhibits. Occasionally the research findings are the result of field investigations and expeditions funded by grants from government agencies and foundations.

To facilitate research, a museum must allocate funds, provide facilities, have a good library, and allow time and support for research, as occurs at the Deutsches Museum in Munich, Milwaukee Public Museum, Museum of Science in Boston, and National Science Museum in Tokyo.

Another form of research is conducted at the Lawrence Hall of Science in Berkeley, which is concerned primarily with research and development in science education. It seeks to improve science curriculums, science education materials, and science teaching techniques.

Most research that takes place at science and technology centers is related to exhibit preparation, exhibit and program evaluation, visitor profiles, and educational programs involving independent study and use of laboratory facilities. These investigations normally do not involve collections to any degree.

Exhibits

Although most exhibits at science centers are constructed for educational purposes, many museums use objects from their collections in exhibits. Depending on the nature and size of the objects, they may be exhibits in themselves or they may be included as part of more comprehensive exhibits.

Those museums with the largest collections—principally in the natural history field—make extensive use of artifacts and specimens, ranging from butterflies to dinosaurs. Collections of plants, minerals, birds, insects, animals, and fossils are invaluable in the preparation of natural history exhibits at such institutions as the Charlotte Nature Museum, National Science Museum in Tokyo, Fernbank Science Center in Atlanta, Science Museum of Minnesota in St. Paul, and Museum of Science and Natural History in St. Louis.

Museums concentrating on the physical and life sciences also utilize objects for some of their exhibits. These artifacts consist of automobiles, machines, locomotives, scientific instruments, airplanes, medical apparatus, and other such hardware. Their number usually is small and the size of the objects generally is large, such as the locomotive ride at the Franklin Institute Science Museum or the U-505 submarine at the Museum of Science and Industry.

In some cases, objects from collections are used to illustrate life in the distant past. They also can be employed to show mankind's progress through the display of old and new equipment in communications, transportation, and other fields.

Teaching

Objects are invaluable in educational programs at science and technology centers, particularly for young children in natural history classes. Nothing can quite compare to having live animals, real fossils, and shiny Indian arrowheads for youngsters to touch and enjoy.

Museum collections also are useful for teenage and adult courses and programs. Spaceships, computers, antique cars, and other objects stored or exhibited at science centers often are included in educational demonstrations and comparisons.

Artifacts and specimens also may be included in teacher's guides, quiz sheets, self-guide tour literature, and supplemental publications provided to schools for museum visitations and classroom study.

Loans

Objects from collections sometimes are made available to other museums, schools, and community organizations on a loan basis. They usually are short-term loans and frequently require insurance, pickup and delivery, and the payment of a fee.

Artifacts sometimes are provided to other museums for use in special exhibitions of limited duration. Such loans may range from a mummy for an Egyptian exhibit to a locomotive for a transportation show.

Objects in school loan programs most often are from the natural history field, such as stuffed animals, sea shells, and fossils, although some loans include instruments, photographs, and textiles.

Science centers also make loans occasionally to community groups for holiday festivities, parades, anniversaries, bank exhibits, and other such purposes.

15

Permanent Exhibits

A Difference in Philosophy

It is largely the exhibit philosophy that separates science and technology centers from traditional museums. Most museums are collection-oriented and have objects of intrinsic value as the focal point of their exhibits. Science centers, on the other hand, minimize collections of objects and emphasize constructed exhibits of a participatory nature.

Traditional museum practices call for a "hands-off" approach, while science center activities are based primarily on a "hands-on" philosophy. Such a basic difference has enormous implications for the development, production, and operation of exhibits at science centers.

It is this difference in exhibit approach that also accounts for the mushrooming popularity of these unconventional science-based museums. The *Illustrated London News* described the Exploratorium in San Francisco as "a participatory museum, or as the Americans with their gift for pithy jargon also term it, a 'hands-on museum.' The exhibits, designed and built by the museum staff, are meant to be manipulated, pushed, pulled, opened, turned, jumped on, and climbed through. There are no priceless treasures in glass cases, no 'do not touch' signs and no guards. The one rule is that bicycle riding among the exhibits is prohibited."[1]

Such a description could be written about most science centers, whether in Philadelphia, Paris, or Singapore. Learning about science, technology, industry, health, and other fields through exhibit participation is the chief ingredient of science and technology centers.

The use of pushbuttons, levers, cranks, and other such participatory exhibit techniques are common at science centers, but they are not the only exhibit methods employed. Live demonstrations, simulated environments, working models, telephone narrations, animated objects, miniatures and enlargements, projected dioramas, videotape monitors, large-screen films, and many other techniques are utilized to attract attention, communicate information, and entertain the museum visitor.

Science and technology centers have broken the museum shackles of the past in their efforts to become more effective instruments of public science education. They have experimented, improvised, and innovated in their exhibits and programs. In the process, they have made museum-going much more popular, productive, and enjoyable. Sometimes, they also have offended museum traditionalists, walked a tight rope with industry, flirted with controversial issues, and gone beyond science in community service.

At the turn of the century, G. Browne Goode of the Smithsonian Institution's U.S. National Museum in Washington argued that museums should have exhibits on ideas rather than merely things.[2] In 1903, Oskar von Miller started a new type of museum—the Deutsches Museum in Munich—based largely on explaining ideas and concepts. But the museum world failed to respond to the calling. In 1936, William K. Gregory of the American Museum of Natural History in New York was still writing about the need for museums of ideas rather than things.[3]

When the exhibit revolution finally occurred in the succeeding decades, science centers were at the forefront of the participatory movement. They placed less emphasis on historic objects and more emphasis on contemporary information. They began to deal with scientific ideas and concepts in an understandable and entertaining way, frequently without the use of artifacts and specimens. They also sought to appeal to the average person with little or no knowledge or interest in science and technology.

G. Ellis Burcaw classified exhibits in two basic ways: according to their purpose or intent, and how the material is organized. When classified by purpose or intent, exhibits can be aesthetic or entertaining; factual; and conceptual. When classed by the way the exhibit material is organized, exhibits usually are systematic—organized according to the similarity of objects and their "genetic" relationship to each other—or ecological—organized according to area, "habitat," or living relationship to each other.[4]

In the science museum field, exhibits at contemporary science and technology centers normally would be classified as entertaining, factual, and conceptual, while exhibits at traditional natural history museums generally would be systematic and ecological. But it is possible for exhibits to be both entertaining and systematic, factual and ecological, or any other combination.

The museum world is changing, as pointed out by Carl E. Guthe in a museum management guide published by the American Association of Museums:

> It is becoming increasingly apparent that the philosophy which motivates exhibit policies in the larger museums is that of creating a hospitable, relaxing environment, in which the visitor is encouraged to identify himself with the exhibits, either by recalling some past personal experience, or by associating what is seen with some current activity or interest. An atmosphere of enjoyment and curiosity is created, which stimulates the visitor to seek further information on the subjects illustrated by the exhibits. Attractively arranged objects, accompanied by brief, accurate, and factual labels, and supplemented by charts, drawings, and photographs, contributed to creating the desired mood on the part of the visitor. Exhibits should not be illustrated textbooks, but rather the settings for a stirring experience, an exciting adventure.[5]

However, it is the exhibit philosophy that largely separates science centers from their museum brethren. By emphasizing ideas and concepts, contemporary subject matter, participatory techniques, and enjoyment, as well as enlightenment, they have helped to make museums more than excursions into the past, storehouses of objects, and places of reverence. In the process, science centers have become effective instruments of popular science education.

Characteristics of an Effective Exhibit

What makes a "good" exhibit? The answer will differ considerably from one type of museum to another and even from one museum professional to another. The criteria for judging exhibits also are changing as museums themselves undergo change.

G. Ellis Burcaw believes a good exhibit, regardless of type, should be safe and secure, visible, catch the eye, look good, hold attention, be worthwhile, and be in good taste.[6]

Robert M. Vogel, curator of mechanical and civil engineering at the Smithsonian Institution's Museum of History and Technology, approaches the subject somewhat differently. For an exhibit to tell its story "effectively," he feels it must attempt to stimulate interest and thought, instruct, furnish a sense of historical development and continuity that is meaningful both to the layman and the expert, and establish the relationship of the subject area to the rest of the world.[7]

Alma S. Wittlin, who has written widely on museums and their exhibit and education functions, has criticized the effectiveness of museum exhibits in communicating information. She said museum exhibits can be grouped under three headings: "underinterpretive," "misinterpretive," and "interpretive."[8] She says that underinterpretive exhibits are descendants of the past when private collections were viewed by a knowledgeable few and did not require explanations. But times have changed, and the viewing public is dismayed by exhibits that are not interpreted adequately. If a museum visitor is unable to find sufficient meaning in what he or she sees, the visitor usually will walk away. Wittlin points out that a study of visitor reactions made in the U.S. Science Pavilion at the 1962 Seattle World's Fair revealed that people spent only forty-five to sixty seconds in front of one of the "information-packed and expensively made displays."[9]

"While the underinterpretive exhibit incorporates a conflict between scholar and layman, the misinterpretive exhibit embodies a conflict between teacher and trader," according to Wittlin. She believes the educational message of many museum exhibits has become blurred by the use of "color," "drama," and "over-designed" techniques. "An exhibit contains two simultaneous messages: the intellectual content and the communication arising from shapes, spaces, lines, colors, and lights, and of course from the interrelationships among all those aspects. The two have to be synchronized and have to support each other," she emphasizes.[10]

Wittlin calls "intellectual overload and sensory monotony" as well as "intellectual deficit and sensory overstimulation" among the most serious hazards of communication by exhibits. Instead of becoming "helpmates of scholars," some exhibit designers "overstep their authority in situations of vanishing restraints."[11]

So what makes a good exhibit? There are almost as many opinions as people speaking and writing about the subject. From a science center standpoint, an effective exhibit is one for which the answers to the following questions are yes: Does it stimulate and hold interest? Does it involve viewers physically or intellectually? Does it interpret ideas or concepts in an understandable manner? Does it result in a pleasurable experience? Does it communicate the message?

It is possible to have a good exhibit without using objects from collections, but most exhibits can be improved through the selective use of artifacts and specimens. When objects are employed, then an additional question should be asked about the exhibit: Does it make effective use of objects to tell the story?

In scientific studies of exhibit effectiveness, another question is raised: Does it meet the predetermined objectives of the exhibit? Proponents of exhibit evaluation argue that every exhibit should have stated objectives and that the finished product should be measured against these goals. Unfortunately, most exhibits are planned, designed, and built without fully stated objectives. The closest they come to such objectives is the "message," which is rationale for having the exhibit.

Watson M. Laetsch, former director of the Lawrence Hall of Science, has been critical of exhibit research and design at all types of museums. He stated:

We still know very little about our audiences and their expectations, and there is no real theory of design (other than what pleases designers). Exhibit design and presentation still mostly follow the textbook model. Visual information prevails, words are the primary medium, and information transfer is the basic goal. This is in spite of one of the few well-known facts of visitor behavior; they read very little. Academia continues to provide the dominant role model for museums, and paradox though it is, communication skills are not highly regarded in academia. A lot is known about the acquisition, care, and feeding of artifacts, and a lot of money is spent in the pursuit of these arts. Very little is known about communication with the public about artifacts and relatively little money is spent in the practice.[12]

Chapter 20 deals with the evaluation of museum exhibits and programs as well as other types of institutional studies. It points out the need for more effective ways of measuring the quality and impact of exhibits. Suffice it to say here that a substantial gap exists between the subjective judgment of exhibit producers and the scientific approach of exhibit evaluators.

The key to any measurement process is the exhibit user—determining how he or she is affected by the exhibit experience. Most exhibit people, however, are more concerned with producing what they believe is a "good" exhibit than with spending additional time and money for evaluative studies that may confirm or refute their intuitive feelings. But an increasing number of museums are taking greater interest in formal evaluation techniques, despite the cost and limitations.

Permanent Exhibits Versus Temporary Exhibitions

A "permanent" or "continuing" exhibit differs from a "temporary" or "traveling" exhibition in several important respects. A permanent exhibit usually is installed in a fixed location for a prolonged period, generally for two or three years or more. In most cases, it also is more comprehensive and participatory and costlier than temporary exhibitions, which are shown for a limited period of time (normally for several months) and sometimes travel to other museums. It is largely the permanent exhibits that determine public attitudes toward a science and technology center.

Exhibit Content

Most permanent exhibits at science and technology centers are concerned with the natural sciences (physics, chemistry, astronomy, geology, biology, engineering, etc.), mathematics, and medicine. These are fields of science and technology that were largely overlooked by traditional museums before the advent of science centers.

The content of the exhibits in a science center should be based principally on the center's objectives and resources and the community's needs and support, which should be determined upon the inception of any science and technology center. Definite choices must be made early about the theme or storyline to be used and the exhibit techniques to be utilized.

Exhibits in science centers usually are contemporary rather than historical, although history may be a part of the story. They also seek to explain scientific principles, technological applications, and the social implications of scientific and technological phenomena. This is most often accomplished through constructed devices that communicate information in a participatory manner.

Most science centers do not have curators to develop the content for exhibits, but they normally have competent staff members to provide the guidance and research for exhibits. Depending on the nature, size, and organization of the museum, the content input can come from the director, education staff, and/or exhibits personnel. When the necessary know-how is not available internally, a science center may hire consultants, use advisory groups, or contract for the exhibit content and design to be handled by an outside firm that specializes in exhibits.

In many ways, an exhibit's content is far more significant that an exhibit's design; yet, from a practical standpoint, exhibit design is just as crucial in attracting the viewer and communicating the information. An ineffective exhibit can result from either poor content or poor design.

Exhibit Design

Exhibit design receives greater attention at science and technology centers than it does at other types of museums. It is the design rather than the objects that most often become the vehicle for conveying the story of an exhibit. The design concept and techniques also have another vital role: they stimulate interest in the exhibit and make the experience enjoyable as well as informative.

Much has been written about the favorable and unfavorable influence of curators and designers on museum exhibits. Writing about natural history museums in 1963, Stephan F. de Borhegyi, then director of the Milwaukee Public Museum and a leader in exhibit innovation, asserted that

Most interpretation is one-sidedly conceived. The curator dreams up an exhibit. He calls in an artist from the art department, and together they make the dream a reality. Only too rarely does the curator consider whether the exhibit will be understood and appealing to the visitor. If he does question its effectiveness, he will likely ask another curator if he understands and approves of it. To make the situation worse, the museum artist tends to design exhibits for the approval of artists in other museums or for designers in the community.

The problem, of course, is that museums are not aimed at the understanding and appreciation of curators and artists. No matter how artistic the layout, how scrupulously accurate the scientific label, if the exhibit does not attract the interest or reach the intellect of the average museum visitor, it is simply wasted time, money, and effort. Visual communication is a form of language. We must not allow it to become a dialect understandable only to our professional colleagues. If we allow this to happen, we are defeating the educational purpose of the museum.[13]

Circumstances have changed somewhat in the last two decades, even among object-oriented natural history museums, but the basic problem still remains the same: the content specialists frequently want to tell too much and/or fail to interpret adequately for the visiting public, and the exhibit designers often produce exhibits that are attractive and/or participatory but are almost meaningless. Science centers always must be on guard against exhibits that are too technical; have too much copy, make use of too many specimens and artifacts, are overdesigned, are too superficial, too commercial, and/or fail to communicate in a meaningful way. A science center staff, however, must be willing to experiment with new exhibit concepts and techniques in an effort to find even more effective ways to interpret the substance of exhibits to the public.

Exhibit designers are storytellers. It is their job to develop design concepts for exhibits within the framework of the museum and the subject matter that will entice, titillate, and leave an imprint on the visiting public. There are no rules that say it is necessary to use objects, arrange the exhibit sequentially, or make use of any "hands-on" devices. The design circumstances will differ with each institution and exhibit.

In 1957, Carl E. Guthe wrote that three basic types of exhibits were being used in small community museums, all based on objects from collections. The first type displayed all materials in the collections in a crowded, uninteresting manner; the second systematically arranged groups of essentially similar objects, often neatly and attractively displayed; and the third and "more rare exhibit policy" subordinated the objects to an overall theme because "they have more meaning if they are used to illustrate principles of association of change or growth in art, history, or science."[14]

In 1975, G. Ellis Burcaw took a somewhat different path in discussing exhibit approaches. He grouped the exhibit designs in four categories: "open storage" approach, in which all objects are placed on display as acquired; "object" approach, a planned educational exhibit based on objects; "idea" approach, a story or idea is presented through the use of objects; and "combined" approach, in which the curator selects both objects and ideas at the same time. Burcaw admitted that it was possible to develop an "idea" exhibit without objects but that it was an "extreme" approach that was a bad practice.[15]

Today most exhibits at science and technology centers are "thematic" in nature, explaining an idea, concept, or story, frequently without the use of objects from collections. The "object" approach is still used, but principally in natural history and other types of more traditional museums.

Among the most common considerations in designing permanent exhibits are the science center's nature, size, and exhibit policies; the exhibit's purpose, budget, and space; the use of objects, participatory techniques, cases, labels, lighting, color, and movement; and concern for city codes, maintenance, security, safety, and the handicapped.

Type of Museum

An exhibit designer must consider a science center's purpose and nature in determining what design is appropriate for such an institution. For example, is the institution's primary objective to preserve and interpret artifacts and/or specimens or to serve as a contemporary science education center? Does the nature of the science center call for greater emphasis on objects or participatory techniques?

Size of Museum

The size of a science center makes a considerable difference in the exhibit design approach. If the museum is small and has few visitors, it may call for a different approach than an exhibit that would go into a large science center with a diverse attendance in the hundreds of thousands or millions. A delicate exhibit may not survive in a metropolitan area, but it may bloom in a small town.

Exhibit Policies

Many science centers have written or unwritten policies regarding exhibits, their nature, scope, construction, and maintenance. These policies, for instance, may require the use of objects, limit the height of exhibits, insist on the use of energy-saving bulbs, prohibit the distribution of literature, or restrict the use of certain exhibit techniques.

Purpose of Exhibit

What is the objective of the exhibit? This is a crucial consideration in designing any exhibit. A designer must have a clear understanding of the exhibit's purpose if he or she is to design the exhibit to convey the message to the visiting public. Many exhibits miss the mark because designers fail to grasp this significant point.

Exhibit Budget

The budget for an exhibit has great importance for an exhibit designer. A $100,000 budget, for example, gives a designer considerably more design flexibility than a $10,000 budget. Yet, if the budget is $10,000, a designer must be able to produce the exhibit within the financial limitations.

Exhibit Space

The size, shape, height, windows, and other physical aspects of the proposed site for an exhibit can have considerable influence on the exhibit design. Most exhibits are designed for a particular space. Sometimes, the location of an exhibit will be changed to accommodate a design, but the reverse occurs more frequently.

Use of Objects

When artifacts and specimens are utilized in exhibits, an exhibit designer must make special provisions for their display, interpretation, and protection. It may be difficult to mix a "hands-on" approach with objects from the museum's collections.

Participatory Techniques

Most science centers want participatory devices in their exhibits, but not just for participation reasons. They believe that "do-it-yourself" techniques can be more interesting and meaningful if designed in an effective manner. A participatory exhibit sometimes is highly entertaining but of little educational value.

Display Cases

An exhibit designer has a wide range of display cases available for his or her use. Should the cases be against the wall, serve as free-standing island units, or be integrated into a participatory unit? Should the objects in the cases be viewed from the front, side, or above? What type of labeling, lighting, and security would be most appropriate for the cases?

Labeling

The explanatory labels in exhibits have changed dramatically over the years. Once scientific in nature, they were long and boring. The objective today is to make them concise, interesting, and interpretive. Readability also is a factor in the selection of label type styles and sizes as well as the placement and lighting of labels.

Lighting

The manner in which light is used in an exhibit can greatly influence the overall mood, the exhibit focus, the readability, the damage to sensitive papers and textiles, and even the museum's energy bill. A designer must decide what type of lighting is best and where it can be placed most advantageously.

Color

Colors and textures can transform a drab exhibit into a sparkling success when used properly by a designer. But they do not have to be gaudy or harsh to be effective. Subdued tones serve just as well under the right circumstances. It is the designer's job to utilize the colors and textures that work best in attracting and holding attention and creating the most conducive environment for learning.

Movement

Studies have shown that color and movement are most effective in attracting visitors to exhibits. A moving object in an exhibit hall always seems to draw people to it first. The public is curious and wants to know what is causing the movement. However, unless the movement is used constructively in the exhibit design, it is merely a passing fancy with little or no educational impact.

City Codes

An exhibit designer must be aware of the city's building, electrical, fire, and health codes to ensure the safety of visitors and staff. Exhibits should be designed to comply with such codes, and contractors should be required to meet city regulations in designing, fabricating, and installing exhibits.

Maintenance

Exhibit maintenance can be costly and difficult when a designer fails to take maintenance into consideration. Exhibits that break down easily; that cannot be serviced readily; that invite vandalism; and that use parts that are difficult to order are the bane of science centers.

Security

In designing an exhibit, a designer must take security needs into consideration. It is necessary to protect valuables, discourage vandalism, and eliminate designs that might result in thefts, assaults, and other crimes. An exhibit should be open for surveillance by a guide or guard; objects of value should be securely enclosed or fastened; and tempting items, such as earphones and projectors, should be protected by the design.

Safety

Sharp objects, glass panels, moving devices, loose wires, and other exhibit elements can be dangerous to both visitors and employees. An exhibit design should minimize such safety obstacles. Otherwise, the result can be an injury and possible liability suit.

Facilities for the Handicapped

With the increased emphasis on access for the handicapped, a growing number of science centers are installing ramped entrances, public elevators, and toilet stalls for the handicapped and presenting special programs for people with physical and mental problems. Some museums also have incorporated special sound, visual, and touch provisions in the exhibits for the hearing and visually impaired and others with handicaps.

Exhibit Techniques

Science and technology centers have pioneered numerous forms of exhibit techniques. The dominating characteristic of science center exhibits is participation—the involvement of the visitor in exhibits—but other types of exhibit techniques also are utilized at science centers: object-based techniques, panels, models, simulations, and audiovisuals.

Participatory Techniques

Exhibit participation can be achieved directly or indirectly. In most cases, it involves physical movement, but it also can be intellectual in nature, with little or no physical interaction. Among the most common participation techniques are activation, question-and-answer games, physical involvement, intellectual stimulation, computers, and live demonstrations.

Activation

The pushing of buttons, lifting of levers, and moving of handles are basic to many exhibits at science centers. Such devices activate bouncing balls, chemical demonstrations, miniature trains, energy experiments, audiovisual presentations, and other exhibit units. They were among the first hands on methods and still remain among the most common and popular.

Activation methods can be ineffective if exhibits are not organized and/or designed properly. The exhibit content may be presented in a way that is too technical, too dull, or too long. In such cases, museum visitors may not wait for completion of the message cycle. A related problem with small children is the pushing of buttons without waiting for the message. This can be minimized through better teacher and school group orientation and guidance in the exhibit halls.

Any science center using activation techniques must have the capability to repair such exhibit units. Because of their popularity, participatory exhibits of all types require constant preventive and corrective maintenance. No institution can afford to develop a reputation for malfunctioning exhibits and out of order signs.

Question-and-Answer Games

Museum visitors enjoy playing games, particularly when they stimulate their curiosity or challenge their intellect. Question-and-answer devices frequently are used for this purpose. They present a series of questions with multiple choice answers. Exhibit viewers usually are asked to select the right answers by pushing the appropriate buttons and then are informed by an illuminated panel whether they made the correct choices. Sometimes the correct answers are given immediately, and other times the viewer must keep trying until he or she makes the right selection.

A question-and-answer game makes it possible to focus on a few important points in an exhibit. It also serves to whet a visitor's appetite for more information about the subject. Such devices have been used to pose questions on dental hygiene, water management, electrical principles, and environmental pollution.

Physical Involvement

The riding of a bicycle to produce energy, the use of a microscope to see minute specimens, and the pulling of ropes in a simple machine demonstration are examples of physical involvement in exhibits.

Physical involvement ranges from merely walking through a simulated environment, such as a coal mine or old-time street, to performing scientific experiments with laboratory equipment, such as oscilloscopes and computers. Some such exhibit techniques require close museum supervision and/or additional liability coverage.

Physical involvement is an effective exhibit technique, but it can be expensive and troublesome from a maintenance standpoint. One science center had to discontinue the public use of microscopes because of damage to slides; another removed a simulated driving exhibit unit because of repeated breakdowns resulting from constant usage by teenagers.

Intellectual Stimulation

Participation can be intellectual as well as physical in nature. For example, the viewing of holographs, mathematical formulas, computer patterns, optical illusions, history walls, and other such exhibit materials can be called mental rather than physical experiences. The Exploratorium in San Francisco has many perception exhibits that fall in this category.

Computers

Computers are being employed in an increasing degree in science center exhibits. In addition to their use in exhibits dealing with the operation of computers, they are utilized to involve the visitor and to communicate information in many other types of exhibits. The Lawrence Hall of Science in Berkeley has about fifty computer terminals that are employed to teach computer use, to conduct research, and to play computer games. Lawrence Hall also operates a computer network that includes nearly sixty schools in the Bay area.

The Des Moines Center of Science and Industry makes use of a computer in explaining the nature of computing and information systems. The Museum of Science and Industry in Chicago has computers that provide visual readouts on food and nutrition, banking and economics, agriculture and the farmer, and aviation and space exploration. Each exhibit has from four to nine computer terminals that function similarly to question-and-answer devices.

Computers have great promise in exhibits. They can retrieve information, serve as a question-and-answer source, animate explanations, and even play musical games or simulate a landing on the moon. They are available on location and remotely; have printed and visual readouts; and their images can be produced in black-and-white or color. They can, however, be costly and require expert maintenance from inside or outside sources.

Live Demonstrations

Live demonstrations of scientific principles and technological applications can be given in exhibits or in theaters. Among the most common demonstrations are those dealing with electricity and magnetism, simple machines, chemstry and cryogenics, sound and acoustics, and light and optics.

At the Palais de la Découverte in Paris, nearly all of the demonstrations are given in the exhibits. The Ontario Science Centre in Toronto gives most of its demonstrations in special theaters. In addition to its exhibits and theaters, the Franklin Institute Science Museum in Philadelphia also utilizes a mobile cart to present demonstrations throughout the building.

In most science center demonstrations, one or two staff members describe the nature of the subject and perform a number of experiments to demonstrate principles, processes, and/or applications. Frequently museum visitors are invited to participate in some of the experiments.

Object-Based Techniques

Artifacts and specimens are used alone and in combination with other exhibit techniques at science and technology centers. Sometimes, however, they are not utilized at all because of the nature of the exhibit or design, or because the institution simply does not have the necessary objects.

As mentioned earlier, three basic approaches are used in exhibiting objects from museum collections: open storage, selective display, and thematic grouping. In open storage, all of the museum's collections are placed on exhibit. Selective display involves showing only some of the collections, while thematic grouping provides for exhibiting those objects related to a particular topic. In general, science centers make the greatest use of thematic groupings in which objects may be unsecured, fastened, enclosed, hung, animated, or presented in a diorama or recreated scene.

Unsecured Objects

Although still common, the placement of an object on exhibit without any protective measures is rapidly disappearing at science centers. It no longer is possible to display unsecured artifacts or specimens at many institutions because of thefts and damage caused by museum visitors. Partly because of the hands-on philosophy, everyone wants to touch objects, regardless of their value. As a result, most unsecured objects in science centers are either too large and tough to steal or damage (such as locomotives) or are excess or worthless objects that are expendable (such as skins and minerals). In such cases, the public often is encouraged to handle the objects.

Fastened Objects

Most artifacts and specimens in exhibits now are either fastened or enclosed for their security. When fastened, an object normally is held in place by one or more wires, screws, or chains affixed from the bottom or rear. The objective is to make it much more difficult for the object to be stolen, particularly if it is a small object that can easily fit into a pocket.

Enclosed Objects

The placing of artifacts and specimens in glass cases or plastic bubbles is the most common method of displaying objects in science centers. Such enclosures are extremely helpful in reducing losses, but it is vital that they cannot be knocked over easily, that secure locks are used, and that screws are not readily accessible to the public. When displaying valuable objects, it also may be necessary to use a security system that sounds an alarm when the enclosure is breached.

Hanging Objects

It sometimes is possible to hang objects in an exhibit hall. The Museum of Science and Industry in Chicago, for example, hangs its airplane collection as a means of attracting attention, providing security, and making exhibit floor space available for other purposes. Before hanging objects, it is essential to determine that the ceiling can safely support the weight of the artifacts. The objects also must be high enough off the floor or be fenced off so that they cannot be pulled down by visitors. The cleaning of hanging objects can be difficult and costly, but it is a periodic necessity.

Animated Objects

Moving objects appeal to museum visitors. Clocks, machines, and vehicles sometimes are animated as a means of attracting public attention and demonstrating their operations.

Dioramas

Dioramas come in two forms: miniature and full size. Artifacts and specimens frequently are used in full-size period rooms and wildlife scenes with realistic surroundings. Period rooms usually are found in history museums, while wildlife dioramas are most common at natural history museums.

Recreated Streets and Villages

Recreated streets and villages often make use of artifacts from collections to give authenticity to the historic scenes. In most cases, however, prized objects are not used for such purposes. The greatest use of recreated scenes is made in open-air historical museums, although they can be found at a number of science centers.

Panel Techniques

Two-dimensional panels are found in nearly every exhibit. They often contain art work and copy designed to entice and inform the visiting public. They help to enhance an exhibit's appearance and to interpret its message. They also are far less costly than most other exhibit methods. Yet, most science centers consider panel techniques to be among the least effective exhibit design tools.

Flat, static panels usually do not attract or hold people's attention although they can be helpful in presenting supplemental information. A few panels generally are necessary to bring an exhibit together, but too many panels can be disasterous. When used alone, panels can become monotonous, boring, and even produce a negative reaction. Therefore, extreme care must be exercised in utilizing panels in exhibits.

Graphic and explanatory panels are among the most common types of exhibit panels. Three less frequently used but more effective panel techniques are the illuminated box panel, animated panel, and the history wall, which give panels an added dimension that can be of considerable interest to museum-goers.

Graphic Panels

The title panel and other panels with graphics and large type fall into this group. They normally are used to identify the exhibit and give it continuity. Most graphic panels are attractive but lack substance.

Explanatory Panels

The storyline of an exhibit usually appears on explanatory panels, which contain copy and/or photographs for interpretive purposes. They can be extremely helpful, but too often the copy is too short or long, the illustrations lack appropriate captions, and the type is too small or low to read.

Illuminated Box Panels

The use of illuminated transparency boxes in graphic and explanatory panels can make a substantial difference in viewer interest, especially if the transparencies are eye-catching. However, such illuminated panels should be more than decoration. They should help tell the story.

Animated Panels

Panels that simulate movement through technomation and similar techniques can be quite effective in explaining a scientific principle, industrial process, or bodily function. Sometimes the panels are set in motion by visitors.

History Wall

A "history wall" is an extension of the panel concept. It usually runs twenty or more feet in length and contains copy and illustrations that trace the development of a field.

Model Techniques

Models serve as a substitute when the real thing is not available or when a principle, process, or operation can be explained better with a three-dimensional reproduction. Models can be replicas, miniatures, enlargements, and working models. Working models were among the first participatory exhibits at the Deutsches Museum in Munich at the turn of the century, and they still are popular. Working models usually are activated by a button or crank, but in some instances they are demonstrated by museum personnel because of their value, fragility, size, or potential danger to the public.

Some models can be built by the museum staff, but the more complex models generally have to be obtained from a model, machine, or exhibit shop with the necessary talents, tools, and materials. Occasionally, a hobbyist will produce a professional-quality model.

Replicas

Vehicles, machines, instruments, and other objects sometimes are reproduced as full-scale replicas when the original does not exist or a science center does not have it but feels a duplicate can be extremely helpful in furthering public understanding and appreciation of the subject. High-quality replicas can be costly and nearly as valuable as the original objects.

Miniatures

Miniature models are used frequently in science and technology centers and range from doll houses to railroad models to small dioramas. They are particularly helpful in dealing with large objects, such as locomotives and machinery. Among the science centers with extensive miniature railways are the Museum of Science and Industry in Chicago, Buhl Planetarium and Institute of Popular Science in Pittsburgh, and Des Moines Center of Science and Industry.

Enlargements

An enlarged DNA molecule, heart, or nuclear model makes it possible to visualize small objects or concepts that otherwise would be difficult to comprehend. A number of science centers, including the Franklin Institute Science Museum, Oregon Museum of Science and Industry, and Museum of Science and Industry in Chicago, use giant walk-through heart models in explaining its structure and function.

Working Models

A model with moving parts can be more illuminating than an original that does not move. Working models can be full-scale replicas, miniatures, enlargements, and even transparent anatomical models, and they can be operated continuously, activated by the exhibit viewer, or demonstrated by museum personnel. Most are activated by a button, lever, or crank. Thus, most working models really are participatory exhibits.

Simulation Techniques

The simulation of an environment can be extremely effective, as evidenced by the popularity of full-scale dioramas, period rooms, recreated streets, and industrial simulations at science and technology centers. Such simulated scenes take museums visitors into jungles, old-time mansions, and coal mines and down yesterday's main streets. They give the public a first-hand impression of the conditions described in the exhibit. Simulation techniques can be expensive, but they also can be the highlight of a museum visit.

Dioramas

Full-scale dioramas are most common in natural history museums, where they normally depict wildlife scenes, but they also are used to show native life and early industrial production. Dioramas usually are enclosed in glass and without sound or activation, but they can be placed behind railings, given sound effects, and even be activated.

Period Rooms

Living rooms, factories, laboratories, and other such settings from other periods in history sometimes are reproduced and equipped with original or replica furniture, clothing, machinery, scientific apparatus, and other materials for the public to view through glass or from behind a railing. Period rooms most often are found in museums with a historical emphasis.

Recreated Streets

The simulation of old street and town scenes can be found at some science centers. Museum visitors are able to walk along the street and look into store windows and period rooms and perhaps even see live craftsmen at work or buy a souvenir or an old-fashioned ice cream soda. Among the museums with such recreated streets are the Milwaukee Public Museum, Center of Science and Industry in Columbus, and the Museum of Science and Industry in Chicago.

Industrial Simulations

The "Coal Mine" at the Museum of Science and Industry typifies walk-through industrial simulations. The Deutsches Museum and

Center of Science and Industry in Columbus also have their versions of mines. Such exhibits that recreate underground and other working conditions usually require guided tours because of their cramped quarters, potential hazards, and the need for explanations and demonstrations. However, such tours are extremely popular and can be a helpful source of revenue.

Audiovisual Techniques

The use of audiovisual techniques has accelerated with the participatory movement. Sound and visual methods can be extremely effective, especially when the quality is excellent, the utilization is appropriate, and the science center is not saturated with audiovisual devices.

Narrations, slides, films, and planetarium projections have been used for many years. Among the new audiovisual techniques are videotapes and videodiscs, talking heads, projected dioramas, Chinese mirrors, multimedia presentations, and super-size films. Some have limited use, while others can be applied in almost any exhibit situation. They can operate continuously or be activated by museum visitors or employees.

Audiovisual techniques have considerable potential, but they also can be costly, troublesome to maintain, and ineffective when they are shallow, too long, or overutilized. Yet, an audiovisual show can be an attention-getter or the highlight of an exhibit—or even a museum visit—when properly produced and employed.

Narrations

A taped message is the most common and least expensive form of audiovisual technique found in science and technology centers. Explanatory narrations can be continuous or be activated by a push-button, sonic system, or lifting of a telephone. Taped messages generally are in a localized area around an exhibit unit or restricted to a single person through the use of pickup phones. In science centers, taped narrations frequently are used to describe something taking place in an exhibit, such as a working model, diorama, film, or slides. The voice, complexity, and length of a narration often are determining factors in their success.

Slides

The use of transparency slides to illustrate points in an exhibit can be a cost-saving and colorful technique. Such slides nearly always are 35 mm. Rear projection of slides is rapidly replacing front projection because the equipment is more compact, not as much of a security risk, and can be operated in a lighted room. In addition to fixed installations, some science centers have mobile rear projection boxes with narration capabilities that can be used in temporary exhibitions and as supplements to permanent exhibits. In using both slide and film techniques, the original transparencies and films rarely should be utilized, and a backup duplicate should be kept for emergencies and to replace faded and damaged slides and films.

Films

In addition to presenting science and other films in a theater, science centers sometimes use short films in exhibits. They usually are 16 mm and frequently are automated and on a continuous loop. Exhibit films are most effective when they are in color, fast-moving, and short in length (less than a minute or two). When they run more than several minutes, seating should be provided. Some science centers are using 70 mm films for large-screen shows in small theater areas within exhibits, such as the "Agrisphere" and "Circus Cinema" at the Museum of Science and Industry.

Planetarium Shows

A planetarium show makes use of audiovisual techniques in telling about the planets, stars, and other aspects of the universe. Planetarium projectors come with a variety of capabilities and now are being supplemented by film and slide projectors, lasers, and "Omnimax" systems that produce space, light-music, and high-fidelity film shows on planetarium domes. The new planetarium techniques have made it possible to use the planetarium facility for more purposes and to produce more revenue for the museum.

Videotapes and Videodiscs

Videotape monitors can be found in an increasing number of exhibits, in many cases replacing slide and film presentations because of their high-quality images and convenient video cassette tapes. A few science centers, such as the Center of Science and Industry, have developed videotape capabilities that also are used for educational, publicity, and other purposes. The Museum of Science and Industry in Chicago utilizes both videotapes and newly emerging videodiscs.

Talking Heads

The projection of a talking image on the head of a mannequin to give a lifelike impression is a relatively new exhibit development. It has been used to project the likenesses of Uncle Sam, George Washington, Christopher Columbus, and Thomas Edison on full-scale figurines. The faces and voices actually come from actors. The technique nearly always stops museum visitors.

Projected Dioramas

Instead of a three-dimensional diorama with a painted background, it now is possible to project a transparency diorama on a shaped surface to give a real-life impression. This technique, pioneered by Richard Rush Studios in Chicago, can be less expensive and faster to produce than the traditional diorama. The projection comes from the front and requires a darkened environment.

Chinese Mirrors

This exhibit technique makes use of trick mirrors to show three-dimensional images such as people, objects, and art work. Illusionary in nature, the mirrors produce lifelike images that capture the attention of the visiting public.

Multimedia Presentations

The term "multimedia" usually means the use of three or more slide and/or film projectors and screens, but it also can involve other audiovisual techniques. Such multi-image shows are used in exhibits to tell a comprehensive story in pictures and to attract and hold the attention of visitors. They usually require a theater setting but also can be projected from the front or back of multiple screens.

Exhibit Development and Production

Most science and technology center exhibits are conceived, designed, constructed, and installed by museum staff members. Many science centers believe they can do a better job than any outside source in interpreting the material. Others feel that it is more economical or that it is a logical extension of the curatorial or developmental process. In such cases, the science center normally has one or more designers or artists, an exhibit workshop, and a team of craftsmen. The Ontario Science Centre has one of the largest exhibit staffs: five scientists and twelve scientific assistants, eleven exhibit designers, and twenty-four craftsmen in the woodshop, ten in the electronics and electrical workshop, and eight in the metal shop. The Center of Science and Industry, Dallas Health and Science Museum, Oregon Museum of Science and Industry, Cleveland Health Education Museum, Exploratorium, and Lawrence Hall of Science also design and build many or all of their own exhibits and contract work from other museums. As an offshoot of its exhibit development program, the Exploratorium has published and sells several "cookbooks" on its exhibits.

Science centers utilize outside contractors for a variety of reasons. Some new or small institutions do not have the necessary capabilities to design and fabricate their own exhibits. However, it is almost possible to produce an "instant museum" by purchasing off-the-shelf health models, duplicating exhibits from other museums, and contracting the services of an exhibit designer/builder who can assist in the planning and fill the gaps with other exhibits.

Some of the larger science and technology centers "farm out" the research, design and/or construction of their permanent exhibits and use their own designers, craftsmen, and shop facilities for planning and coordinating the exhibit program, producing temporary exhibitions, and maintaining exhibits. These centers find it difficult to justify a large and diverse exhibit staff for a limited number of major exhibits requiring specialized skills and techniques. They also want to utilize the best design and fabrication talents available and want their exhibits to look different from each other and appeal to the visiting public with a distinctive approach. In many instances, the exhibits are funded by companies, government agencies, and health organizations that want the best, and it is easier to account for costs on an out-of-pocket basis.

At the Museum of Science and Industry, all design and construction contracts and payments can be made directly by the sponsor or through the museum, but they all must be approved and authorized by the museum. The Franklin Institute Science Museum, Singapore Science Centre, California Museum of Science and Industry, Science Museum in Tokyo, Museum of Science in Boston, and New York Hall of Science also use outside designers and builders for their permanent exhibits.

Exhibit Sponsorship

One of the most costly items in any museum budget is the exhibit program. A temporary exhibition normally ranges in cost from a few hundred dollars to $50,000. Most permanent exhibits, on the other hand, begin at $50,000 and run $1 million or more because of their greater size, sturdier construction, and greater complexity. Permanent exhibits that utilize participatory, audiovisual, and other sophisticated techniques cost from $50 to $300 per square foot, with the average being about $150. Thus, a 1,000-square-foot exhibit would require about $150,000. In addition, annual maintenance may cost $1,000 to $10,000.

Many science and technology centers try to obtain contributions, grants, and/or sponsors for both temporary and permanent exhibits. In the United States, grants for temporary exhibitions are available from the National Endowment for the Arts, National Endowment for the Humanities, National Science Foundation, Institute of Museum Services, and state arts and humanities councils. It also is possible to solicit nominal contributions from members, trustees, companies, and foundations to cover the expense of building a temporary exhibition. As might be expected, it is much more difficult to raise the larger amounts of money needed for building and maintaining permanent exhibits.

Most science centers that have sponsored exhibits obtain such support from companies and trade associations based in the region; local or state branches of medical, health, and professional organizations; and sometimes local, state, and federal government agencies. It is extremely difficult to obtain sponsored exhibits from outside the immediate region unless the museum is large, has an annual attendance in the millions, and draws visitors from throughout the nation.

Although sponsored exhibits can be a bonanza, they also can be a problem. A science center must have a policy on the types of sponsored exhibits it will accept and the conditions under which it will accept them. A center also must maintain complete control over the content, design, and construction of sponsored exhibits. Otherwise, they can become commercial instead of educational; impugn the integrity of the museum; violate electrical and other city codes; result in costly maintenance; cause the energy bill to skyrocket; and/or alienate visitors, other contributors, and segments of the community. If an exhibit sponsor will not agree to modify a questionable exhibit, it nearly always is in the best interests of the museum to turn down the offer. It may be a short-term loss, but it nearly always is a long-term gain.

A science center must decide whether the subject matter of a sponsored exhibit is compatible with its purposes. The storyline may be historic or contemporary, but it must be broad based and educational. A sponsor's equipment and products generally can be used as examples in developing the theme, but they should not be presented competitively. In return for supporting an exhibit, a sponsor usually is identified on a modest plaque or sign at the entrance and exit.

"What these companies get for their money is plenty of exposure and the opportunity for some low-key image building," according to *Business Week*.[16] This may not be enough to justify a major investment by some museum exhibit sponsors, but many of the large corporations, trade associations, government agencies, professional societies, and health organizations are convinced that sponsored educational exhibits at science centers represent an effective use of institutional or public service funds.

Among the sponsors of exhibits at American science centers are International Business Machines Corporation, Eastman Kodak Company, General Motors Corporation, Standard Oil Company (Indiana), United Airlines, International Harvester Company, Esmark Inc., American Iron and Steel Institute, U.S. Department of Energy, Environmental Protection Agency, U.S. Air Force, National Institutes of Health, American Medical Association, American Dental Association, National Foundation, American Cancer Society, and Arthritis and Rheumatism Foundation.

Critics have charged that some sponsored exhibits are biased, commercial, and/or instruments of propaganda. It has been said that certain sponsored exhibits mislead the public by presenting an incomplete, distorted, or even inaccurate picture in some fields. Typical of such comments is the one by Howard Learner of the

Center for Science in Public Interest: "Instead of education, people get propaganda as some museums limply deal with technological superficialities, rather than underlying causes. . . . Without assertive curatorial control by the museum, the exhibits can result in glossy, image-building advertisements, at best, and blatant corporate propaganda masquerading as 'education' at worst."[17]

Some sponsored exhibits do have promotional messages—to make greater use of nuclear power, to stop smoking or using drugs, to support the military services, to conserve energy and use alternative forms of energy, to reduce environmental pollution, to eat more nutritional and less caloric foods, to control population growth, to improve dental hygiene, to continue the space program, and other such objectives. Some exhibits of this type are considered more acceptable than others. For example, most sponsored exhibits designed to further better health are considered to be in the public interest.

Unlike traditional museums, science centers are dealing primarily with the present and future, rather than the past. In seeking to further public understanding of contemporary science and technology, science centers cannot avoid sensitive issues and still fulfill their missions. But every science center has an obligation to society to present factual information and unbiased interpretations in its exhibits, regardless of the sponsorship.

In general, the overall impact of sponsored exhibits has been extremely favorable. Outside support has made it possible for science centers to develop new exhibit techniques and present significant exhibits in many scientific, technological, industrial, and medical fields that formerly were ignored by museums.

The key, of course, is strong-willed museum management that will not sacrifice integrity for dollars—the same ingredient needed in dealing with donations of art works presented to art museums, historical monuments suggested to history museums, and collections of specimens offered to natural history museums.

Exhibit Maintenance

Effective and prompt exhibit maintenance is a necessity for every science and technology center. Any hands-on exhibit, regardless of how well it is conceived or constructed, needs both preventive and corrective maintenance from time to time. Unless a science center has a sound exhibit maintenance program, a flood of out of order signs and disenchanted visitors can undermine the institution's attendance, support, and existence.

The maintenance of participatory exhibits is quite different from the care of collections, although some of the same personnel and techniques may be involved. At a small science center, one or two people may have the responsibility for all exhibit and collection maintenance. At larger institutions, as many as fifty may be concerned with various aspects of such maintenance. In addition to the internal maintenance staff, some science centers also make use of outside specialists, particularly for machinery, electronic devices, and computers.

In addition to qualified personnel, a science center needs specialized equipment, adequate work space, and appropriate storage. Too often museum planners fail to allow enough space for working on exhibit construction and repair and storing exhibit materials and collections. It is especially important to store exhibit materials and collections separately. Almost any large, dry, and clean space with adequate access can be used to store exhibit devices, panels, lights, and other such exhibit materials.

Exhibit maintenance varies considerably from one science center to another, but regardless of the personnel or procedures, exhibit maintenance usually includes the following.

Cleaning

All exhibits must be cleaned periodically or they will deteriorate or become unsightly. Cleaning usually consists of removing fingerprints, wiping off dust, and picking up debris. Such work can be done routinely by janitors. However, the cleaning of valuable objects on display and in storage may require personnel with special training.

Conservation

Exhibits and collections also need "touching up" occasionally because of scratches, worn paint or type, and cracking. Normally such work is performed by an artist at science centers or a conservation specialist at collection-oriented museums.

Woodworking

Sometimes it becomes necessary to repair, modify, or rebuild the structure of an exhibit. This usually requires one or more carpenters and woodworking equipment. At most science centers the carpenters who help build exhibits also handle such maintenance.

Mechanical Repair

Most participatory exhibits have mechanical parts that need attention periodically. A large science center may need a machinist and machine shop to make mechanical repairs; a museum with relatively few participatory units may utilize a handyman.

Electrical Work

Virtually every museum has one or more electricians or other knowledgeable employees to handle normal electrical work for the building and exhibits. When the work becomes extensive or complicated, an outside contractor may be called to assist.

Electronic Repair

The increasing complexity of exhibits generally requires inside or outside personnel capable of repairing the electronic circuitry of malfunctioning exhibits. Under certain circumstances, an electrician with further training can perform this task. Increasingly, it requires electronic technicians who specialize in such work.

Audiovisual Work

Taped narrations, slides, films, videotapes, and multimedia presentations, which can be found at almost every science center, require preventive and corrective maintenance by trained personnel in a well-equipped shop. To minimize maintenance problems, most science centers attempt to standardize projection, videotape, and multimedia equipment.

Computer Maintenance

Few science centers are capable of maintaining computers used in exhibits. Therefore, a service contract with the supplier or other external source usually is the best answer. If the exhibit is sponsored by a company, such maintenance may be provided by service personnel from the firm's computer center.

Because the maintenance of exhibits is expensive, some museums have exhibit maintenance agreements with sponsors of exhibits. These contracts, which generally run from three to five years, provide for the performance of certain services in return for the payment of an annual maintenance fee. These services may cover only the janitorial and technical maintenance of the exhibit or they may encompass guide, security, and/or demonstration services and insurance coverage. Normally, such agreements do not include major repairs, such as replacing a compressor or carpeting, which must be supplied by the exhibitor when needed.

Two approaches are used most often in determining the amount of the annual maintenance fee. Some science centers base the charge on the square footage of the exhibit. For example, a flat charge of $10 per square foot is used at several institutions. Other science centers attempt to estimate the amount of maintenance and other services involved and apply an annual charge based on experiences with comparable exhibits. In other words, the larger and more complex the exhibit, the higher the maintenance charge. Most exhibit maintenance fees are in the $1,000 to $10,000 range, although some are $25,000 or more.

16

Temporary and Traveling Exhibitions

Temporary Exhibitions

Temporary exhibitions differ from permanent exhibits in several important aspects. As the name implies, they are of limited duration, normally being shown from several weeks to a number of months rather than for years. They generally have a theme and focus on a narrower or more specialized segment of a field and normally are smaller, less substantial, and frequently less costly than permanent exhibits.

Temporary exhibitions can be time-consuming, difficult to assemble, and even costly. They often are developed on an overload basis and interfere with other museum responsibilities; some are minuscule in size and mediocre in quality. So, it properly might be asked, "Why bother with them?" The answer, of course, is that all museums, and especially science centers, *need* temporary exhibitions to build attendance, raise funds, obtain publicity, attract members, fill voids in permanent exhibits, inform the public on important or interesting topics, provide a community service, and sometimes just fill space. Temporary exhibitions also are needed as an outlet for an institution's collections and the creative and professional zeal of staff members.

Thus, both the homegrown and traveling varieties of temporary exhibitions play an important role in the life and success of a museum. They give people an excuse for returning to an institution and for helping to support its activities. They increase a museum's vitality and provide a more comprehensive service. They also serve as a change of pace and new challenge for the exhibits, curatorial, and education staffs.

Types of Exhibitions

Locally produced temporary exhibitions take numerous forms. Many are originated, researched, designed, fabricated, and installed by the science center's exhibits, curatorial, and/or education staffs. Some are conceived internally but are produced by outside suppliers on a contract basis. Still others come to the museum from community, corporate, and other sources, such as science fairs, holiday displays, and technical societies, health organizations, and industrial firms.

Temporary exhibitions produced by the museum staff may be based on objects from collections, subjects not covered adequately in permanent exhibits, or topics of current interest to the staff, members, and/or the general public. The Cranbrook Institute of Science in Bloomfield Hills, Michigan, for example, prepared an exhibition entitled "The Last Days of Pompeii" that made use of artifacts, volcanic specimens, a scale model, photographs, and other materials from its collections and elsewhere. The Lawrence Hall of Science produced a show called "Beauty in Nature" that united several temporary exhibitions with two permanent exhibits, a silk-screen interpretation of Apollo Mission photographs and a collection of exceptional mineral specimens.

More frequently, science and technology centers do not make use of artifacts and specimens or minimize their role in temporary exhibitions. Another approach to temporary exhibitions was a series of quarterly exhibits on timely science advances, issues, and problems at the Musuem of Science and Industry. The same panel mounting system, question-and-answer device, audiovisual equipment, and display cases were used for all four exhibits, but the panel copy, illustrations, participatory questions, films, and objects changed with the exhibit topics—space probes, genetic engineering, energy, and nuclear power.

Local temporary exhibitions frequently are panel shows largely because of budget constraints. It also takes valuable staff time to research, design, and construct a three-dimensional, participatory exhibition, and the investment may not be justified for a limited showing. Temporary exhibitions provided by local sources normally are for annual events, such as student science fairs, health-oriented exhibits, rock shows, and holiday festivals, or one-time events related to anniversaries, conferences, and subjects of special interest to a technical society, the school system, a company, or an ethnic group. The Oregon Museum of Science and Industry has about fifteen annual exhibitions presented by professional, trade, hobby, and other groups. They usually are scheduled for a week or two and cover such subjects as minerals, aviation, reptiles, typography, crafts, beekeeping, bottle collecting, wool and other fibers, dairy goats, native plants, and photography.

Funding Sources

The most common obstacle to locally produced temporary exhibitions is money. Since most temporary exhibitions are optional and funds are limited, a science center tends to commit its resources to necessary activities such as patching the roof, recruiting members, updating permanent exhibits, and offering educational programs. As a result, relatively little funding is available from an institution's budget for temporary shows, despite their potential benefits.

To overcome this financial dilemma, most science centers seek to obtain funds and exhibits from outside the museum. They apply for grants, attempt to find sponsors for exhibits, and seek to convince community sources of the value of presenting temporary exhibitions. They may go to local and state arts councils, private foundations, the federal endowments, corporations, health organizations, professional societies, and individuals for funds or exhibits. For example, the Museum of Science in Boston obtained a temporary exhibition on the history of the snapshot from a business member of the museum. The Oregon Archaeological Society organizes a show at the Oregon Museum of Science and Industry each year. The Continental Bank and the *Chicago Defender,* a black newspaper, provide the funds and personnel for an annual "Black Esthetics Festival" that includes a major art exhibition at the Chicago museum.

Funds are available for good temporary exhibition ideas, but it requires considerable effort—and frequently lead time—to obtain the support. It is necessary to develop the exhibit concept somewhat, to explain its significance, and to point out why a particular source should be interested in the proposed exhibit. To obtain funds from government agencies and corporate and private foundations, a museum director, curator, exhibits head, or educator must fill out application forms and wait for months before receiving a decision.

In seeking funds for temporary exhibitions, a science center must appeal to a prospect's civic pride or try to match a proposed exhibitions's subject or purpose with a possible donor's specific interests. Most foundations have grant guidelines, arts councils and other government agencies restrict funding to certain areas, and companies tend to limit their giving to activities related to product lines or communities in which they have facilities.

Concept and Design

The concept and design of temporary exhibitions are similar to those of permanent exhibits. The main differences are the construction materials and the design flair. Because temporary exhibitions have a shorter life and/or must be lightweight for travel, they usually make use of less durable but more portable materials. The brief life of a temporary show also allows designers to try new and flashier techniques to attract attention and separate them from the permanent exhibits, without commiting an institution to such a long-term design policy.

In his booklet on museum work, G. Ellis Burcaw said the principal difference between a permanent and temporary exhibit, aside from the lifespan, is that a temporary display ordinarily employs a greater measure of what he calls "show business,"[1] which he defined as "entertainment with a flair; organized, "slick" gratification of the senses. It is drama and opera—and going over Niagara Falls in a barrel. It is conspicuous, noisy, sometimes gaudy and superficial, but universally appealing. Show business may employ the fine arts, but its aim is at mass audience. It uses sound, color, movement, excitement, animals, sex, suspense, and strange and skillful behavior to entertain and to produce a favorable response in its audience. . . ."[2]

In the museum world, showmanship is the use of exhibit techniques to attract and hold the visiting public, to generate interest and entertain, and to create a favorable impression. It may involve the use of colors, movement, sounds, artwork, lights, participation, working models, live demonstrations, or multimedia shows.

Unfortunately, most locally produced temporary exhibitions fall far short by "show business" standards. The culprit can be the budget and/or staff imagination. Science centers and other museums rarely have the necessary funds to take full advantage of these techniques. The show business approach is more common in major traveling exhibitions.

However, it still is possible to incorporate eye-catching elements with objects, panels, and other exhibit materials, despite limited funds. The judicious use of colors, lights, graphics, and other low-cost techniques almost always can convert a drab, uninteresting exhibition into a high-interest, effective show.

Although design innovations are to be encouraged, a museum must be careful not to let the exhibit design determine or dominate the exhibit content or message. An exhibit concept should be the result of a close working relationship among curatorial, education, and design staff members, with the design being an implementation of the concept and exhibit purpose. A design staff should always avoid cases that are too crowded with objects; panels that have too much copy; lighting that damages documents, textiles, and other sensitive materials; labeling that is too small and difficult to read; slide and film presentations that are too long or loud; participatory devices that do not work reliably; and exposed artifacts that are not protected adequately.

A science center's exhibits staff also should try to provide professional guidance and assistance to local schools, companies, professional societies, health organizations, hobbyists, and ethnic groups that donate temporary exhibitions to science centers but without assuming responsibility for organizing, presenting, or funding community exhibitions. It should be remembered that whatever appears in a museum reflects that institution. Therefore, it must be selective in offering its facilities for and accepting outside exhibitions. It always must retain control of what exhibitions are presented and what goes into exhibitions, and strive for the highest possible quality under the circumstances.

Development and Production

The development and production of a temporary exhibition may take place internally and/or externally, depending on the size and capabilities of the staff, the nature and scope of the exhibition, and the timetable and budget.

In most science and technology centers, temporary exhibitions are developed and produced by the exhibits staff with the assistance of the curatorial and/or education departments. On large or complex exhibitions, museums sometimes will use outside designers and/or fabricators. Institutions that have internal design and production talents and facilities, such as the Ontario Science Centre, Oregon Museum of Science and Industry, and Center of Science and Industry, nearly always develop and construct their temporary exhibitions. However, some science centers, such as the Museum of Science and Industry, Museum of Science in Boston, and California Museum of Science and Industry, will hire outside exhibit designers and/or builders for major temporary shows.

Working with outside design or fabrication contractors can be difficult and expensive, but it also can be extremely helpful. Much depends on the groundwork laid by the museum staff and the way in which outside help is utilized. It is possible, for example, to use contractors to develop only the overall concept or design. Instead of having all the construction "farmed out," a few stock or custom exhibit units may be all that is needed to supplement existing cases, panels, audiovisual equipment, and interactive devices. Many of the same exhibit units can be used for several temporary exhibitions, particularly those that are relatively simple.

Interpretation and Promotion

The interpretation and promotion of homegrown temporary exhibitions frequently are overlooked. In many instances explanatory literature is not published and volunteers are not briefed on the exhibits. Other times the promotion amounts to a single news release with no photographs, posters, interviews, price previews, feature stories, or any of the other publicity possibilities.

It is not enough to say to the public "the exhibit is ready—come to see it." A science center should try to interpret an exhibit through informed guide service, collateral literature, slides and films, live demonstrations, lectures and seminars, teacher workshops, and perhaps even courses and field trips. It also should attempt to promote interest in the exhibition in the media, through direct mail, and in the museum's periodic publications. In addition, an institution should try to capitalize on such an event by having an opening reception for members and donors, sponsoring a conference on the exhibit topic, and/or having an open house for teachers or parties interested in the subject.

A temporary exhibition should be more than a space-filler or an annual going-through-the-motions exhibit that lacks interest and excitement. It requires time, money, and space to prepare and should produce a return for the museum and the community. Almost any exhibit worth doing also is worth interpreting and promoting.

Traveling Exhibitions

Traveling exhibitions have increased in number and popularity as museums have developed the need for more temporary exhibitions. They range from simple panel exhibits to elaborate shows with printed catalogs, educational materials, and promotional kits. They can be two-dimensional or three-dimensional, static or participatory, and inexpensive or costly. The come in all shapes, sizes, and packages. The biggest problem is the shortage of good science exhibitions, especially of the participatory type.

Originating and Receiving

Science and technology centers are both originators and users of traveling exhibitions. Many traveling shows are the result of locally produced temporary exhibitions that were converted to circulating exhibits. Others are produced specifically to travel. On the other hand, some science centers are strictly users of traveling exhibitions. They lack the capabilities, funds, and/or interest to develop exhibitions for circulating to other institutions.

Producing a traveling exhibition is quite different from developing a permanent exhibit or even a local temporary show, primarily in the way it is built. A traveling exhibition must be lightweight, easy to assemble, require little maintenance, and be flexible for a variety of configurations. Original objects often cannot be circulated for fear of damage. It is necessary to have special shipping crates, extra insurance, instructions for unpacking and operating, educational and promotional packets, and other materials that may not be used in local shows.

One of the most frequent problems is the attempt to circulate an exhibit that was not built to travel. Such an effort nearly always ends in disaster with high maintenance costs and general dissatisfaction. Some museums see traveling exhibitions as a way to make money. Actually, it nearly always costs more than the revenues produced because of breakdowns, travel, and unforeseen difficulties.

Science centers that circulate their own exhibits rather than use a traveling exhibition service usually are affiliated with state or federal agencies and receive government subsidies for their activities. In Calcutta, for instance, the Birla Industrial and Technological Museum circulates three traveling exhibit vans throughout the region as an arm of India's National Council of Science Museums. The American Museum of Science and Energy operates a traveling van and exhibit program on energy for the U.S. Department of Energy. The province-funded Ontario Science Centre circulates its "Science Circus" and other traveling exhibits to all parts of Ontario. From Richmond, the Science Museum of Virginia sends its mobile van throughout the state to inform people about solar energy and other science topics.

Traveling exhibitions that are not funded by the government usually are simple panel shows, exhibits whose travel is funded by a grant, exhibits sponsored by a number of museums, or those for which the rental fee is substantial. For example, the traveling exhibition "The Design Science of R. Buckminster Fuller" was developed and circulated by the Museum of Science and Industry and funded in part by the National Endowment for the Arts. In 1973–1974, science centers in Philadelphia, San Francisco, and Chicago worked with the Center for Advanced Visual Studies at the Massachusetts Institute of Technology on a traveling art-science exhibition. At the other extreme, the Museum of Holography in New York covers its expenses for a traveling exhibition on holography by charging a rental fee of $1,250 per week, with a four-week minimum.

Most traveling exhibitions are handled by fourteen nonprofit circulating services and a handful of profit agencies (see list in appendix). These organizations generally have full-time, experienced professionals who are knowledgeable in the circulating and booking of traveling exhibitions. They simplify life for the originating museums and occasionally produce a financial return.

Museums that make use of traveling exhibitions have somewhat different problems from the originating institutions. First, of course, is the matter of exhibit space. Far too many science centers are limited in what circulating shows they can receive because of the lack of sufficient or appropriate space. Most small museums have 1,000 or less square feet when the minimum should be 3,000.

Only a few of the larger science centers have 5,000 or more square feet of temporary exhibit space. The Museum of Science and Industry has a wing of its building for temporary exhibitions and special programs that includes 27,000 square feet of exhibit space, an auditorium, and two theaters. In Boston, the science museum uses an elegant hall and two medium-sized spaces for temporary shows. The California Museum of Science and Industry has acquired a large surplus armory for its temporary exhibitions, education programs, and space exhibit.

It should be pointed out that although traveling exhibitions are helpful in attracting interest and support, they are not a panacea for a museum's ills. A science center must have a base of permanent exhibits, educational offerings, and local temporary exhibitions to obtain sustained community involvement and financial backing.

Circulating Services

The first nonprofit traveling exhibitions service in the United States was founded in 1909. By 1977, there were fourteen such agencies and others in formative stages. They offer nearly 500 traveling exhibitions and book some 2,000 showings a year seen by about twelve million people each year.

In addition to the nonprofit agencies, there are a number of traveling exhibitions services operated for profit, such as Ruder & Finn Fine Arts, a division of a New York-based public relations firm, and Van Arsdale Associates Inc., a circulating service based in Alexandria, Virginia. Many of the Ruder & Finn traveling exhibitions are sponsored by industrial companies and are offshoots of corporate public relations programs handled by the firm.

The traveling services came into being in response to a need to make available worthwhile temporary exhibitions at a reasonable cost. Originally concerned primarily with art exhibits and museums, they have expanded into areas such as history and science and are serving a wider range of users—libraries, schools, colleges, banks, shopping centers, and all types of museums.

Traveling exhibition agencies serve a dual purpose. They are a source of temporary exhibitions for museums and other organizations and a vehicle through which corporations, government agencies, and trade associations can provide circulating exhibitions. Some of the traveling exhibitions are free; others require the payment of a rental fee and/or payment of one-way shipping expenses.

Because of the increasing interest in traveling exhibitions, a workshop was held on the subject in Washington, D.C., in 1977. It was organized by the Association of Science-Technology Centers and funded largely by the National Endowment for the Arts. The workshop report stated that

Traveling exhibitions services function as brokers and facilitators as well as organizers and distributors of temporary exhibitions. They supply exhibitions from small panel exhibits to elaborate three-dimensional shows.

Their products are used to fill exhibit space, build attendance, raise funds, and serve the public by dispensing knowledge. Traveling exhibitions also are used by sponsors as an instrument to disseminate information and to influence public opinion.[3]

In 1977, I conducted a mail survey of the fourteen American non-profit traveling exhibitions services. The results showed that wide differences exist among the various agencies. The services range from two-person shops and serious deficits to twenty-six-person staffs and sizable surpluses.

Art and photographic exhibitions are the most common exhibits circulated by the agencies. Almost all of the services organize some of their exhibitions, but most circulate exhibits supplied by artists, art collectors, galleries, art dealers, museums, corporations, and/or governments.

Only three of the traveling services offer exhibitions relating to science and technology: the Smithsonian Institution Traveling Exhibition Service, Association of Science-Technology Centers Traveling Exhibition Service, and Oak Ridge Associated Universities Museum Division Traveling Programs. The Smithsonian service (commonly known as SITES) is the largest circulating agency, offering more than 200 exhibits and booking over 1,000 showings a year. The ASTC Traveling Exhibition Service came into being in 1974 to meet the need for science-oriented exhibitions. The Oak

Ridge service, administered through the American Museum of Science and Energy, is concerned primarily with energy exhibits circulated for the U.S. Department of Energy.

The largest share of income for traveling exhibitions services comes from rental fees, ranging from zero to 83 percent and averaging nearly 39 percent of the budget. Other major sources of income are federal grants and contracts from government agencies, foundations, corporations, and arts councils as well as private contributions. SITES exhibition fees range from $100 to $15,000.

The principal need of nonprofit traveling services is additional funding. Other needs cited in the survey included better exhibitions, more personnel, less costly and better shipping, and additional sponsors of exhibitions.

Types of Exhibitions

Circulating exhibits seldom consist of a single object, but they frequently have many objects—some valuable originals, some replicas, and some worthless objects. The simplest traveling exhibition may be a suitcase of objects or a box of panels. These usually are circulated in a community rather than long distances. They are inexpensive, easy to assemble, and direct in approach. Sometimes they are accompanied by volunteers, docents, or staff members, who may explain or demonstrate parts of the exhibit.

Another common type of circulating exhibit is the two-dimensional panel show, which may include paintings, photographs, sketches, graphics, and copy panels. Such exhibitions are found most often in the art field. They are least effective in science when dealing with ideas and concepts. Science panel shows have presented a selection of scanning electron microscope photographs; "Transportation in Switzerland," a panel exhibit prepared by the Swiss Transport Museum in Lucerne; and "Future of the Oceans," a pictorial look at the promise of the seas by the Canadian government.

Three-dimensional exhibitions with participatory elements are among the most effective and popular traveling science shows, but they usually are the most expensive to produce and circulate. Such exhibits require a greater investment in developing, maintaining, and shipping sensitive exhibit devices, but they usually are worth it. The public enjoys such hands-on interactive units. The "Think

Metric" exhibition developed jointly by the Oregon Museum of Science and Industry in Portland and the Center of Science and Industry in Columbus typifies the value of such exhibits. It was funded by the U.S. Bureau of Standards and circulated by ASTC. Another effective participatory exhibit was "Illusions in Art, Science, and Nature," developed and built in England. Unfortunately, it could not withstand public use and soon disintegrated.

The mobile van approach is used by a number of science centers, including the American Museum of Science and Energy, Birla Industrial and Technological Museum, Science Museum of Virginia, California Museum of Science and Industry, and Ontario Science Centre. In some instances the vans are walk-through exhibit halls. Other times some or all of the exhibit units are removed for display outside the van or in a nearby structure.

Financing and Budget

The budgets of traveling exhibitions vary from a few thousand dollars to over $1 million. The financial needs are affected by the size, nature, and schedule of the exhibit. Whatever the cost of producing a show, it is virtually impossible to pay for it through exhibit fees. Therefore, except for simple panel exhibitions, it nearly always is necessary to have a grant, contribution, subsidy, or sponsor for a traveling exhibition.

Among the typical sources of income for traveling exhibitions are rental fees from users, catalog and poster sales, slide and other educational material purchases, and occasionally the sale of prints, replicas, and souvenirs. Exhibition fees usually range from several hundred to several thousand dollars, although they can reach $100,000 or more for major shows.

In developing a traveling exhibition budget it is necessary to include all the expenses and not just the cost of producing the exhibit. Among the other expenses are packing, shipping, maintenance, insurance, printed materials, educational kits, photographs and slides, news releases, and office expenses, such as correspondence, postage, telephone, and photocopy services.

The institutional cost for using a traveling exhibition is considerably less, but it also can be costly. The local budget should include such expenses as the exhibit rental fee, one-way shipping, in-house insurance, extra labor and equipment, promotional materials, and opening reception costs. It frequently is possible to find a local sponsor for a traveling exhibition to cover all or part of these costs. The patron may be a company, foundation, or individual. Sometimes a special admission fee for the traveling show may be appropriate and help with the financing.

Research and Storyline

The core of any traveling exhibition is the content. What is the storyline or message? What objects, photographs, diagrams, and other materials should be part of the exhibit? What should be the theme of the show?

These types of questions should be answered largely by the curator, researcher, educator, or person—inside or outside a museum—who is charged with the responsibility of organizing the exhibition. In a collection-oriented institution, the content source usually is the curator. In a typical science center, it can be almost anyone—including an outsider hired to research the subject.

"The most important things for the researcher to remember are purpose and meaning. The basic data must be organized to carry out the purpose of the exhibition," Barbara Tyler and Victoria Dickenson pointed out in their handbook on traveling exhibitions. "Once the storyline has been researched and the artifacts selected, the contours should lead naturally to a certain type of exhibition design."[4]

Traveling exhibitions consisting largely of objects are still common in the art, history, and natural history fields. The great success of the Tutankhamen and Pompeii traveling shows testifies to their popularity when the objects are significant and the exhibitions are well financed and promoted. In science and technology, objects are less frequent because many exhibits are concerned with ideas, concepts, and contemporary topics. Therefore, the emphasis is on interpretation. As Tyler and Dickenson stated: "Artifacts, specimens, and works of art may be the bones of an exhibition, but interpretation is the flesh."[5]

Competent content research should be a necessary ingredient of every exhibit, whether temporary or permanent, stationary or traveling. The "facts" are needed to develop a storyline, theme, or point of view. Before a design concept can be produced, a designer also must know what kinds of objects, photographs, participatory devices, and other materials are available to illustrate the storyline.

In determining the exhibition theme, every effort should be made to make it appealing to the general public. A circulating exhibit that is too technical or esoteric normally will fail at the box office and backfire on both the originating and receiving institutions. It also should be "tightly conceived" to take into consideration the appropriate size of the exhibit, fragility of the work, and capacity of the proposed booking institutions, as explained in the Western Association of Art Museums' traveling exhibitions workbook.[6]

Design and Fabrication

Traveling exhibitions have somewhat different design and fabrication requirements than permanent and temporary exhibits at the local level. Because circulating exhibits must withstand all the rigors of travel, they must be conceived and produced to be lightweight, compact, durable, secure, flexible, and relatively easy to pack, ship, install, and maintain, but without sacrificing "show business."

Designers have considerable freedom with traveling exhibitions, but they still must design the exhibit to follow the storyline, stay within the budget, and meet the basic traveling criteria. The design of a circulating exhibit also must take into consideration such factors as the differing spaces in which the exhibit will be shown; the physical dimensions of the viewers; the security and safety of artifacts and other valuables; and the legibility and readability of the labels, panel copy, and graphics. A typical traveling show ranges from 1,000 to 2,000 square feet, or about 350 linear feet. Most often it is designed for a rectangular room with nine-foot ceilings. But science centers have both smaller and larger temporary exhibitions facilities.

Shabtay Levy, director of exhibits at the Oregon Museum of Science and Industry in Portland, divides traveling exhibitions into five categories: traveling art shows containing two- or three-dimensional original arts; artifacts, specimens, and life-size mockups; informative displays containing only copy and graphics; large exhibition and installations usually housing multimedia equipment; and hands-on exhibits.

At a traveling exhibition workshop, Levy said panel and multimedia shows are frequently ineffective, wasteful, and/or unimaginative. "Why bother to take a book and put it up on the walls?" he asked, when studies indicate that the public spends little time reading the massive copy on panels. As for elaborate multislide and synchronized tape narration installations, he pointed out it would be far less expensive and probably just as effective to supply a museum with a set of slides and tapes for use in an existing environment. Levy believes the most effective and exciting type of traveling exhibition is one that makes use of participatory techniques.[7]

William Brown, coordinator of traveling programs at the American Museum of Science and Energy, feels that many problems associated with traveling exhibitions originate in their design. To avoid "frustrating and expensive operational problems that are inherent in a poor design," he suggests that the following questions be considered in the design process: Will the exhibit fit through the door? Will the sponsor need a crane to get the exhibit off the truck? Will the design require a team of expert craftsmen to decipher the instructions and install the exhibit? Will the exhibit survive the trauma of Mr. Peabody's seventh grade biology class? Will technical wizards be required to maintain the exhibit? Is the exhibit a hazard to human health?[8]

As can be seen from the foregoing, designing a traveling exhibit involves much more than an appealing design. Virtually every facet of a traveling show has a bearing on the design, and the design has an impact on almost everything associated with a traveling exhibit.

Shipping and Maintenance

Registration, scheduling, booking, packing, shipping, insurance, installation, and maintenance are among the operational details of traveling exhibitions. Although not as visible as the exhibit content or design, many of these behind-the-scenes aspects are just as vital to the success of a circulating show.

Emily Dyer, registrar for the Smithsonian Institution Traveling Exhibition Service, listed the three basic concerns of her office as safety, travelability, and security of exhibits. Among the steps in handling an exhibition are analytical consideration of the proposed exhibit; organization of the exhibition; collection, inspection, and conservation of exhibit materials; valuation of works; insurance, shipping, and security arrangements; and other such matters.[9]

The scheduling and booking of exhibitions require extreme care.

Whenever possible, scheduling should be a natural progression of stops rather than a costly zigzagging pattern across the nation or continent. It should be known at bookings that the exhibit will fit properly. Sufficient time should be allocated between showings, ranging from several weeks to a month. The movement of exhibits over weekends and holidays should be avoided as well as the shipping of delicate exhibitions during the winter when extreme shifts in temperature can cause damage. Bookings should be confirmed in writing, with all responsibilities clearly defined.

The packing and shipping aspects of a traveling exhibition are most critical. Regardless of how well arranged, packing and shipping are seldom free of problems. Sturdy crates or boxes should be used for packing cases, and objects and exhibit materials should be protected in custom-shaped styrofoam or ethafoam chambers or placed among such materials as kinpack, styrofoam pellets or noodles, bubble-pack, or acid-free tissue paper. Whenever possible, crates should be light and small enough to be lifted by two people and moved through a standard single door. For larger exhibits, however, a forklift and double doors frequently are necessary. Most traveling exhibits are shipped by truck or plane, and a few circulating services have their own vans. Much of the damage to exhibitions results from poor packing, failure to secure crates in trucks, and careless handling in loading and unloading.

Insurance is mandatory for any traveling exhibition. Wall-to-wall fine arts insurance coverage usually is recommended for the circulator, although receiving institutions frequently insure traveling exhibits while they are on the museum premises. Huntington T. Block, president of an insurance firm that specializes in museums, said he looks for the following in insuring traveling exhibitions: professionalism of the people handling the exhibit, adequate packing and handling, adequate transportation, place of temporary storage, and possible subrogation.[10]

The installation or mounting of an exhibit usually is done by the exhibits staff of the receiving institution. However, on some large, delicate, or expensive shows, one or more supervisory personnel will accompany the exhibit to oversee installation. It is vital that instructions and diagrams for installing purposes are clear and concise. An exhibition is most vulnerable to theft during the installation period when objects and exhibit pieces are scattered around the exhibit hall. Steps should be taken to make the exhibit room secure during this period or to see that all valuable and essential parts are locked in crates or secure storage areas until installed.

The maintenance of traveling exhibitions is a serious problem in the science field. Unlike most circulating shows, science-oriented exhibitions often have three-dimensional, participatory exhibits with moving parts that need adjustment or repair. Sometimes the maintenance work can be performed by the host museum; other times it is necessary to send a maintenance specialist to the exhibit or return the exhibit unit to the fabricator.

Interpretive Materials

Every traveling exhibit should have some collateral materials for interpretive and educational purposes. The basic need usually is a descriptive leaflet for free distribution or a comprehensive catalog for sale. In addition, there may be slides, films, videotapes, books, bibliographies, suggested lectures, educational kits, and live performances to supplement a circulating exhibition. But few traveling exhibitions offer more than a few of the possibilities, and most museums do not utilize what is available.

The concept of an exhibition catalog has changed over the years. At one time a scholarly document with a compilation of technical data, specialized articles, and other information primarily for professionals in the field, catalogs have become more valuable to the average person viewing an exhibit or refreshing his or her memory later. Most catalogs now focus on an exhibition's highlights, with photographs and brief descriptions of the principal objects or sections as well as an overview of the nature, purpose, and significance of the exhibit. A "giveaway" leaflet usually is a smaller and less expensive version of the catalog, with a few panels of copy and illustrations giving background information on the show.

The use of audiovisual techniques, such as slides, films, and videotapes, can be extremely helpful from an interpretive standpoint. Such materials sometimes are included as part of an exhibit or are made available by the exhibit circulator. Science centers frequently will use such materials as part of complimentary programs—illustrated lectures, science demonstrations, seminars, and film series.

Suggested talks, books, bibliographies, and educational kits, which contain these and other materials, can be useful in conducted tours, gallery lectures, courses, and other educational programs. However, such extensive educational materials usually are too expensive to be made available as part of a normal traveling exhibition. Many museums also do not take full advantage of these interpretive opportunities.

In a study of the Smithsonian Institution Traveling Exhibition Service's educational materials and needs, Antonio Diez found a great market for all types of interpretive tools in connection with traveling exhibits. Museums indicated their principal needs were for bibliographies, visual aids, and catalogs or descriptive information for docents and the public.[11]

Promotion and Publicity

Traveling exhibitions are exceptional promotional and publicity vehicles. They can be used for increasing attendance, raising funds, attracting members, and for other promotional purposes. At the same time, traveling shows can be the source of extensive media coverage and a reason for special mailings, events, and articles in museum publications.

A science center must consider the merits of each traveling show on an individual basis. It should be careful not to give equal promotion and publicity to each exhibition, since they differ in quality and promotional value. Although it cannot afford to be indiscriminate without misleading its friends and the press, a museum should be alert to the promotional and publicity possibilities of each show.

For important or popular traveling exhibitions, most museums will hold an opening function for members, local dignitaries, and special-interest groups. This affair may be a luncheon, reception, or dinner with speakers, ribbon-cutting ceremony, entertainment, demonstration, and/or audiovisual presentation. When a traveling exhibit is sponsored, the presenting company often will provide the funds for such an opening event. When there is no tour sponsor, a museum will often convince a local firm or individual to underwrite the opening.

Most traveling exhibitions come with stock news releases, a catalog or descriptive leaflet, and a few photographs that can be adapted for local use. It also may be possible to obtain slides, films, and videotapes for use on television public service announcements and news and interview programs. Some traveling shows also have posters for local imprinting, although many museums prefer to design their own. For significant circulating exhibitions, it also may be appropriate to hold a press preview of the exhibit; produce a special mailer for schools, members, and/or friends; arrange for outdoor signs or streetlight banners; or offer admission discounts through local merchants.

Traveling exhibitions should receive the full promotional and publicity support of science centers, according to Eileen Reynolds, director of audience development at the Franklin Institute Science Museum. "Museums need traveling exhibitions to develop new audiences, attract public attention, and help fill voids in their offerings."[12]

Evaluation

Evaluation has become an increasing concern of circulators, sponsors, and users of traveling exhibitions. It no longer is acceptable in some circles to use the attendance figures or to weigh the newspaper clippings to determine the success or failure of a traveling show. However, despite the desire for scientific evidence of effectiveness, exhibit evaluation practices are still largely subjective.

Nearly all traveling exhibit services have a followup evaluation form, but it is primarily statistical or subjective in nature. In addition to attendance figures and newspaper clippings, such reports usually deal with radio and television transcripts, advertisements, school group attendance, supplemental programming, letters received on exhibit, and the opinions of museum officials.

Traveling services and museums conduct relatively little evaluation that is scientifically valid because of the cost, the questionable value, and reservations about survey techniques. Watson M. Laetsch, former director of the Lawrence Hall of Science, believes that it is the attitudes of the museum professionals, rather than money or techniques, that prevent the greater use of evaluation in the exhibit field. "Evaluation can be ego damaging," he asserts.[13]

Lois-ellin Datta, assistant director of the Education and Work Group, National Institute of Education, said traveling exhibitions have an educational outreach that is surpassed only by public television. "Traveling exhibits are the only educational vehicle that is tried out in different settings and institutions and on different types of audiences," she pointed out. She feels exhibit evaluation is "difficult but possible."[14]

A more common view is presented by John Drabik, former director of exhibits at the Museum of Science in Boston: "You can't evaluate exhibits with traditional techniques. You learn the value of exhibits through experience. Most studies use academic models that are not applicable to museum operations. Evaluation studies do not measure the total experience and transfer later in life." Instead of measuring "what concepts are retained," the studies should be concerned with "what precepts and insights result."[15]

The question of exhibit and program evaluation is considered in greater detail in chapter 20. The measurement of learning and effectiveness is difficult in any field, and especially in the free-flowing environment of a science center, which differs considerably from the controlled classroom situation.

17

Educational Programming

A New Educational Dimension

Unlike many traditional museums, the principal thrust of most science and technology centers is education. Virtually all of a science center's undertakings, whether they are collections, exhibits, educational programs, membership activities, or community services, are aimed at furthering public education in science and technology. Yet science centers generally are not recognized as educational institutions by the government, educators, and even some museum people.

S. Dillon Ripley, secretary of the Smithsonian Institution, has asserted that

. . . Museum visitors tend to feel, vaguely perhaps, that a museum is an educational experience if not an educational institution. Most museum workers tend to cling staunchly to the idea that a museum is an educational institution even though it does not give degrees.

To me, a museum as such, a building containing collections, and with exhibits which are open to the public, is a center for exposure rather than for education. Education—that is, pedagogy, as it is largely thought of today—exists as a function of teaching. It is to a large extent a didactic process with the teacher saying, "Now hear this."[1]

Whether a science center is an educational institution in a definitional sense is academic. Its activities frequently are educational in nature and its impact on learning often is far greater than formal educational systems.

Frank Oppenheimer, director of the Exploratorium in San Francisco, believes that museums are an important avenue for educational "sightseeing." "In performing this role," he said, "they will relieve the schools and universities of a responsibility which schools are not designed to fulfill. Schools cannot easily provide an adequate vehicle for sightseeing. Sightseeing from a classroom situation usually resembles sightseeing from a train. Courses rush to their destination so that passengers can make their connections. The great vistas that rush by, the people and towns along the way, never become part of the viewer's experience."[2]

Museums are different from schools in Oppenheimer's opinion. "No one ever flunks a museum, one museum is not a prerequisite for the next. People do not list the museums they have attended on a job application form. Museums are thus free of many of the tensions which can make education unbearable and ineffective in the schools."[3]

"In contrast to classrooms, museums provide a reversible, deflectable, three-dimensional form of education," Oppenheimer stated. They are voluntary, participatory, entertaining, and necessary "artificially created environments" for learning. Despite this vital role in the educational process, museums have become "orphans of the educational endeavor," he explained.[4]

Among the factors for this situation, according to Oppenheimer, are that museums have not uniformly considered themselves as educational institutions and have not creatively developed appropriate forms of pedagogy; they frequently consider themselves primarily as repositories of objects; many have allowed themselves to be used for glorifying rather than for instructing; some have become more concerned with things than with people; others have not stimulated multisensory contact with the exhibits; and, most importantly, they appear as "play" rather than "work" in the public mind.[5]

Although most activities of a science and technology center are considered educational in nature, the responsibilities of the education staff or department, headed by a director, curator, or coordinator of education, usually are more restrictive. They normally are focused on relations with the schools; class visit arrangements; instructional activities, such as science demonstrations, classes, and field trips; and specific educational services involving collections, exhibits, publications, films, and teachers.

The possibilities for educational programs and services are almost endless. A 1976 survey of science centers by the Association of Science-Technology Centers showed that of the twenty-seven institutions responding 100 percent provide one or more educational services to the public; 93 percent work with schools and other organizations; 89 percent offer one or more services to teachers; and 75 percent work with schools in outreach programs. Among the educational programs and services to schools, teachers, private organizations, and the general public were exhibits, demonstrations, lectures, films, guidebooks, quiz sheets, science kits, and free literature on exhibits.[6]

In summarizing the report, Lee Kimche, then executive director of ASTC, wrote that "Contemporary museums of science and technology . . . are playing an increasingly important role in meeting the educational needs of society. While traditional museums tend to emphasize collections and research, today's science museums are dedicated to furthering the public's appreciation of science and technology through informal exhibit and other learning techniques."[7]

Informal education is education that permits people to approach it on their own terms. "In other words," according to Kimche, "people bring their own agendas to the learning environment, where they explore their special interests in participatory ways. It is perhaps in such an environment that science can be learned best, not only because people learn in different ways, but because informal learning environments allow different channels for learning."[8]

The rising educational demands on science museums present challenges and added responsibilities, she stated. "Museums are beginning to perform functions traditionally assumed to be within the domain of schools. Basically, science centers do not want to serve as surrogate schools; rather, they want to present education in an informal setting that supplements the classroom situation." The ideal situation, she said, is for "the schools and museum to work in collaboration, with schools presenting a formal curriculum complemented by hands-on, object-oriented, participatory exhibits that are available at the science centers."[9]

C. G. Screven, professor of psychology at the University of Wisconsin-Milwaukee and a leading museum reseacher, said, "As a learning environment, the museum provides an alternative place for something called 'education' to take place. The museum as a place for education, in fact, may have some unique advantages over more formalized public education for persons of all types and ages." [10]

"Potentially at least," he pointed out, "the museum is an exciting alternative to conventional education. Museums have no classrooms, no coercive forces, no grades. The visitor is in an exploratory situation, moving about at his own pace and on his own terms. Unlike formal schools, the museum is basically a 'non-word' world of things and experiences presented in real-life proportions. The museum should serve as an ideal learning environment for inviting inquiry, questioning, and constructive practice in investigatory behaviors." [11]

The objective of any science center director and education staff should be to capitalize on this opportunity to make learning more effective and meaningful. By utilizing the appropriate techniques in this conducive environment, science centers bring a new dimension to the educational process.

Educational Activities

The educational activities of science and technology centers vary greatly, from exhibit interpretation to extensive programs of demonstrations, classes, field trips, school services, and outreach programming. How deeply a science center becomes involved in educational activities depends largely on its institutional objectives and resources and the community's educational needs.

The educational work of science centers can be grouped into three categories: basic educational activities, school and outreach services, and other educational programs. Most educational activities take place in the first category.

Basic Educational Activities

Among the basic educational activities offered by science centers are exhibit interpretation, science demonstrations, lessons and workshops, courses, lectures and films, field trips and tours, planetarium shows, and library services.

Exhibit Interpretation

The interpretation of exhibits is one of the most common educational functions in science centers. The education staff frequently is involved in the planning of exhibits from an educational standpoint, in preparing exhibit labels and explanatory literature, and in conducting exhibit tours and producing self-guide tour materials. Although about half of the science centers have guided tours, they are decreasing because of personnel costs and the difficulty in keeping tour groups together in a participatory environment. The Boston Museum of Science instead has volunteer guides stationed throughout the building to answer questions, discuss exhibit content, and demonstrate materials.

Teachers also are playing a more important role in conducting guided tours for their classes. Some science centers, such as the Franklin Institute Science Museum in Philadelphia, Ontario Science Centre in Toronto, and Museum of Science and Industry in Chicago, publish teacher's guides, quiz sheets, and self-guide tour materials to facilitate school group touring.

The Center of Science and Industry in Columbus, Ohio, the Rochester Museum and Science Center in Rochester, New York, and other institutions provide an orientation program for school groups upon their arrival, pointing out the highlights and making suggestions for touring and studying the exhibits. Some science centers, such as the Ontario Science Centre, Museum of Science and Industry, and American Museum of Science and Energy in Oak Ridge, Tennessee, have special "orientation rooms" or "pits" to assist in handling and orienting visiting groups. In addition to its School Orientation Center, the Chicago museum has a Visitors' Center that gives a multimedia tour of the exhibits in six languages.

Science Demonstrations

About one-third of the science centers provide live demonstrations of scientific principles and applications as part of their educational programs. Some of the demonstrations are given in exhibit areas, while others are offered in auditoriums, theaters, and minitheaters. The demonstrations usually are presented by guides, volunteers, and/or education staff members at regularly scheduled intervals and/or when requested by school or other groups. Most demonstrations last from fifteen to thirty minutes, although some run longer. The Des Moines Center of Science and Industry, for instance, has lengthened its demonstrations from thirty to forty-five minutes to allow for greater flexibility in programming and more discussion.

The Palais de la Découverte in Paris has one of the most extensive demonstration programs, with a demonstration counter in nearly every exhibit hall. In Munich, the Deutsches Museum has built a number of minitheaters adjacent to new exhibits for demonstration purposes. In addition to fixed demonstrations, the Franklin Institute Science Museum has a portable demonstration cart that is wheeled about the museum for impromptu presentations in hallways and exhibit rooms.

The Ontario Science Centre has three main demonstration areas for laser, electricity, and chemistry/cryogenics programs and eleven smaller sites for demonstrations given upon request in such fields as printing meteorology, the human body, and electron microscopy. In Chicago, the Museum of Science and Industry makes use of both exhibit areas and theaters in presenting five regular and fifteen requested demonstrations in such fields as chemistry, physics, electricity, biology, simple machines, and electromagnetic spectrum.

Lessons and Workshops

A step beyond demonstrations are lessons and workshops. They usually are one- to two-hour classroom or laboratory sessions on topics closely related to school curriculums. Unlike most demonstrations, they are more formal in nature and frequently require the payment of a nominal fee by the students or schools.

The Boston Museum of Science has fifteen "school programs" for different grade levels. They are grouped as introductory, biology, chemistry, physical science, and planetarium programs. The Cranbrook Institute of Science in Bloomfield Hills, Michigan, offers ten "museum lessons" that cover such areas as ecology, underwater life, and Michigan geological history. The Fort Worth Museum of Science and History has fifty school programs based largely on objects from the museum's collections.

Workshops are more participatory than lessons. Youngsters have an opportunity to explore and experiment under controlled conditions. The Center of Science and Industry in Columbus, for example, has four "Make-It Take-It" workshops in such fields as chicken embryology, astronomy, optical illusions, and the human life process. The Lawrence Hall of Science offers hands-on family and children's workshops on elementary computer programming, microscopic bio-art, animal behavior, human health, and other subjects.

Among the most active science centers in the lesson and workshop field are the health-oriented museums. The Cleveland Health Education Museum works with the school system to provide instruction in such areas as dental care, sex education, and drug abuse. The Kansas Health Museum in Halstead makes available formal lessons in nine subject areas, including senses, heredity, and human growth and development. The Robert Crown Center for Health Education in Hinsdale, Illinois, presents educational programs for students, adults, and teachers in four major fields: general health education, family living/sex education, drug abuse prevention education, and environmental education/human ecology. The center uses exhibits and electronic devices in a classroom-theater setting to serve more than 150,000 people each year.

Courses

Educational courses for children and/or adults are offered by three out of four science centers, according to the Association of Science-Technology Centers educational survey.[12] In more than one-third of the cases, they are credit courses presented in cooperation with a school system, college, or university. About half of the science centers provide opportunities for independent study. "Courses" differ from "lessons" in that they cover a subject in greater depth, involve a number of sessions, and require payment of a larger fee. Many courses also are presented when the museum normally is closed.

Courses at science centers take many forms. They are presented for children, adults, and/or teachers, and are offered in the evenings, on Saturday, and during the summer. In general, the purpose is to provide opportunities for science education without duplicating local school or college programming. In some instances, the courses are not related to science but are designed to fill adult education needs in the community.

The Boston science museum has six categories of courses. The "Discovery Series" is for young children from kindergarten through third grade and consists of twelve sessions on alternate Saturdays.

Twelve one-hour lecture-demonstrations for grades four through eight are offered on the other Saturdays as part of the "Explorer Series." A "Teachers' Series" on science topics, involving ten two-hour sessions, is offered for elementary teachers on a credit basis through Boston University. The three other course categories—"Young People's Courses," "Adult Astronomy Series," and "Navigation Series"—are related to the planetarium program.

In Los Angeles, the California Museum of Science and Industry presents an "Exploring Science on Saturdays" program for children and teachers. Twenty-five courses, each two hours on four consecutive Saturdays, are offered on such subjects as sea animals, energy, rockets, chemistry, photography, computers, and geology. Saturday workshops also are presented for teachers, with graduate credit available through the University of California at Los Angeles Extension Division.

The Exploratorium in San Francisco has a summer class program that involves courses in optics, electricity, electronics, computers, and aerodynamics. The classes meet for two hours on six days. The Des Moines Center of Science and Industry's summer program provides ten courses for four grade groupings and for hobbyists. The children's courses are given from Monday through Thursday for one hour and fifteen minutes and deal with animals, meteorology, plants, geology, computers, and biology. The adult courses are for two hours on Tuesday and Thursday evenings over a three-week period and cover such topics as photography and fiber dyeing.

The Oregon Museum of Science and Industry offers evening, Saturday, and summer courses in many fields, including biology, mathematics, chemistry, arts, crafts, and dramatics. It also has an "Energy Center," which seeks to familiarize the public with issues, concerns, and alternatives regarding the use of all forms of energy. Evening classes and tours are among the activities.

The Franklin Institute Science Museum emphasizes laboratory-oriented courses in its "Discovery Science Workshops." Twenty workshops aimed at various grade levels and adults cover such fields as astronomy, microscopy, criminology, chemistry, photography, electricity, and amateur radio. The two-hour classes are held on weekdays and Saturdays and last from one to two months.

The Rochester Museum and Science Center offers a wide range of adult and children's courses—many outside the field of science—

through its "School of Science and Man," which was made possible by a $750,000 grant from the late Mrs. Caroline Gannett, wife of the founder of the Gannett newspaper chain. Typical courses include candlemaking, Chinese cooking, dancing, needlepoint, cross-country skiing, photography, wine tasting, and astronomy. The length of the class sessions and courses vary.

Lectures and Films

Nearly all science centers have lectures and films as part of their educational programs. Lectures sometimes take the form of seminars, symposiums, and public forums, while films may include slide presentations, multimedia shows, and other audiovisual productions of an educational nature. Some lectures and films are presented as part of the daily operation of science centers; others are scheduled only in the evenings, on weekends, and/or on an irregular basis. Lectures and films may be free and open to everyone, restricted to members, or have an admission charge.

The Lawrence Hall of Science has a weekly series of lectures and films. The Thursday evening free lectures are given by leading scientists and engineers on space probes, solar energy, ecology, and other timely science topics. Films for young children, teenagers, and adults are offered several times each Saturday and Sunday. Although some of the films are general in nature, most are science related.

In Boston, the Museum of Science offers science films on Friday nights as part of the museum admission charge and a series of free Wednesday evening lectures for adults. The lecture series—funded with the assistance of the Lowell Institute—is given in cooperation with the Harvard-Smithsonian Center for Astrophysics and may be taken for Harvard Extension School credit. A recent series dealt with "Life in the Universe."

The Cranbrook Institute of Science presents a monthly evening lecture series and a semimonthly Saturday afternoon film series on science subjects. The lectures are open to the public for a fee, with a reduced rate for members. The films are free to members and available to others as part of the museum admission charge.

The Ontario Science Centre has daily films and weekly lectures and illustrated talks as well as various special programs. The free daily films are mostly children's films scheduled largely for afternoons and evenings. In addition, an "Ontario Film Theatre" presents general films from throughout the world Tuesday through Friday, with a "Senior Citizens' Matinee" every Wednesday. The public lectures and illustrated talks usually are given on Sunday afternoons and feature scientists, explorers, and others. In addition, special presentations are made periodically for high school and college students on weekday afternoons.

A different approach is used at the Museum of Science and Industry in Chicago. Rather than special lectures and films, the museum shows motion pictures continuously in eleven minitheaters as part of permanent exhibits. The educational films are concerned with such subjects as agriculture, communications, energy, steel-making, air transportation, and health. Evening seminars and public forums are held from time to time on medical advances, science issues, and technology and human values.

Field Trips and Tours

Many types of field trips and tours are available at science and technology centers. Field trips usually are of short duration—from a few hours to several days—and involve trips within the immediate region. Organized tours, which often are arranged through travel agents take museum groups to distant points for periods ranging from a week to a month or more. Most of these traveling activities are part of educational programs, although some are primarily extensions of membership programs. In all cases, advance registration and fees are necessary.

The Museum of Science and Industry has monthly field trips to local industrial facilities and research laboratories. Among the all-day trips have been visits to such places as steel mills, auto assembly plants, railroad yards, water treatment plants, banks, agricultural and locomotive equipment plants, petroleum refineries, newspaper plants, and government and industrial laboratories.

The Charlotte Nature Museum offers both field trips and tours. For example, it has sponsored a "Wildacres Weekend" for families, which included two days at camp, hiking, crafts, lapidary demonstrations, and a visit to an emerald mine. The museum also sponsors an annual "Environmental Tour" with the local school system and 4-H clubs. In 1979, students in the eighth through twelfth grades toured the northeastern United States and Canada.

At the Franklin Institute Science Museum, field trips have included a trip to New York to attend a performance of *Madame Butterfly* at the Metropolitan Opera. Among the tours have been a two-week trip to Iceland to see lava fields, volcanos, geothermal power sources, glaciers, birds, and unusual vegetation. The tour was limited to twenty to thirty-five people, cost $1,600, and was conducted by a knowledgeable member and a licensed Icelandic guide-lecturer.

The Children's Museum of Indianapolis organizes minitours for members and their families. In the fall of 1978, for example, two tours were offered: a four-day trip to St. Louis, Hannibal, and the Dickson Mounds Museum, and a three-day trip to New York, which included a Broadway play, sightseeing, shopping, and museum-going. The Rochester Museum and Science Center also has similar tours for members, such as a recent three-day sight-seeing trip to San Antonio.

Planetarium Shows

About half of the science centers have planetariums that offer "sky shows." Some "theaters of the stars" also have astronomical and space exhibits or offer such audiovisual extravaganzas as "Omnimax" domed-screen films or "Laserium" laser light and music shows. Some planetarium shows are included in the price of admission to a museum, while others call for the payment of a fee. The more elaborate film and light shows always have a separate admission charge, whether presented during the day or evening, and sometimes are a substantial source of revenue.

The Fort Worth Museum of Science and History and the Dallas Health and Science Museum offer planetarium programs to school groups and the public. Fort Worth's planetarium uses a Spitz A3P projector, while the Dallas instrument is a Minolta MS-8. The sky shows range from a look at the stars to space exploration.

In Atlanta, the Fernbank Science Center has an observatory as well as a planetarium with a Ziess Mark V projector. In addition to regular planetarium shows, the Charlotte Nature Museum produces such unusual programs as "Captain Cosmos and the Space Puzzles Affair" and presents Sunday evening tours of the planets, stars, constellations, and other sights in the sky over Charlotte.

Boston's Museum of Science offers thirty-five-minute lecture-demonstrations in the planetarium several times daily, with the show changing periodically. The museum also has star-gazing programs beginning at 3 P.M. each day.

The Strasenburgh Planetarium at the Rochester Museum and Science Center has one of the most comprehensive facilities and outstanding records of special programs. Between 1969 and 1979, it produced 11 theater programs, 10 music programs, 4 multimedia shows, 1 dance program, and 5 light shows. The 306 performances were attended by 54,891 people or an average of 179 people each performance, which was 75 percent capacity. In addition, the planetarium presented 810 "Laserium" and "Laserock" light shows—produced by Laser Images of Van Nuys, California—that were attended by 126,599 people.[13] (Multicolored laser beams are projected inside the planetarium dome to the accompaniment of music in such light shows.)

Among the science center planetariums equipped with the 70-mm Omnimax projection systems with 180-degree fish-eye lenses are the Rueben H. Fleet Space Theater and Science Center in San Diego, Detroit Science Center, Science Museum of Minnesota in St. Paul, and Alfa Cultural Center in Monterrey, Mexico. Developed by Imax Systems Corporation of Cambridge, Ontario, the high-fidelity motion pictures projected on the inside of a planetarium dome have proved to be extremely popular. During the first four years of operation in San Diego, the Omnimax show attracted an average of 400,000 annually.[14] In the first four months of operation, an average of 1,760 people (nearly six times the capacity of the 300-seat theater) came to see the film "Genesis" at the Science Museum of Minnesota's new "Omni-theater."[15]

Library Services

Libraries can be found in about four-fifths of the science centers, but many are not open to the public. Most of the libraries are rather small, restricted to science and technology, and used primarily by staff members. About one-third have one or more special collections on such diverse subjects as astronomy, hydroelectric power, and transportation.

A 1975 survey of science center libraries revealed that they were generally "underfinanced, understaffed, and underutilized."[16] Of the thirty-two institutions with libraries in the survey, nine reported that no funds were spent for library personnel, thirteen made annual purchases of books and periodicals totaling less than $3,000, and eight spent nothing on library equipment. Although nine libraries did not use audiovisual and other media, slides were available at sixteen museums, records at eleven, cassettes at nine, films at nine, filmstrips at eight, super-8 loops at six, microfilms at five, microfiche at five, and videotape recordings at five.[17]

Although twelve of the libraries offered special services to children, such as book fairs, classes, reference books, workshops, and assistance on science fairs and other projects, few had books or audiovisual materials especially for children. For instance, eight libraries had no science books for children, nine had fewer than 100, eight had 101–1,000, and only two had more than 1,000. Only six of the libraries had audiovisual materials designed for children— mostly films and filmstrips.[18]

Despite these discouraging figures, some excellent libraries can be found at certain science centers, such as the Franklin Institute Science Museum in Philadelphia, Museum of Science in Boston, and Lawrence Hall of Science in Berkeley. Franklin Institute's library has more than 250,000 volumes and receives 4,000 periodicals regularly from throughout the world. The Boston museum has a 33,000-volume library, which is said to be an inviting place for "browsing, borrowing, and doing research."[19]

School and Outreach Services

In addition to offering a variety of educational programs at science centers, many institutions provide helpful services in the schools, communities, and regions they serve. These include educational publications, audiovisual materials, loan materials, science demonstrations, traveling exhibits, enrichment programs, and in-service teacher programs.

Educational Publications

Every science and technology center sends publications to schools in its area. Among the most common printed materials used to communicate with and assist schools are newsletters, calendars of events, magazines, teachers' guides, program notices, catalogs, quiz sheets, self-guide tour materials, and literature on specific exhibits and subjects.

Among the science centers that publish monthly or bimonthly

newsletters and calendars are the Center of Science and Industry in Columbus, Charlotte Nature Museum, California Museum of Science and Industry, Rochester Science Museum and Science Center, Maryland Science Center, Oregon Museum of Science and Industry, Ontario Science Centre, and Museum of Science and Industry in Chicago.

A number of science centers produce professional journals that are mailed to schools and others, such as *Kultur & Technik* by the Deutsches Museum in Munich, *The Journal of the Franklin Institute* in Philadelphia, and *Revue du Palais de la Découverte* by the Paris science center, which feature technical and/or educational articles. Some institutions have more modest bimonthly or quarterly magazines dealing with their activities. They include the *Franklin Institute News, The Exploratorium,* and the *Science Centre Bulletin* in Singapore.

Many science centers have annual program notices and instructions for school group visits that are mailed to schools, such as the annual "Guide to OMSI Happenings" from the Oregon Museum of Science and Industry, the "School and Group Visits" folder from the science museum in Boston, and the "COSI School and Group Visitation Guide" from the Center of Science and Industry in Columbus. They are designed to inform teachers about educational programs, procedures, and costs.

Teacher's guides, quiz sheets, background information, and/or self-guide tour materials are available from science centers to assist teachers in planning school group visits. One of the most extensive arrays of useful teaching materials is published by the Ontario Science Centre in Toronto, which includes "Teacher's Guide" leaflets, "Exhibit Enquiry Series" quiz sheets, and "Reference Series" background sheets on exhibits.

The American Museum of Science and Energy and the Lawrence Hall of Science have developed a series of printed experiments in energy and health for use in schools and museums. The "Science Activities in Energy" packets—funded by the U.S. Department of Energy—illustrate principles and problems related to various forms of energy and to their development, use, and conservation. They are designed primarily for fourth, fifth, and sixth graders. The Lawrence Hall's "Health Activities Project" materials, made possible through grants from the Robert Wood Johnson Foundation, seek to inform fifth through eighth graders of their own health and safety through participatory approaches.

Audiovisual Materials

Films, slides, filmstrips, records, tapes, and videotapes are among the audiovisual materials available from some science and technology centers. Nearly half of the science centers offer one or more of the materials, with science films being the most frequent service.

The Fernbank Science Center offers various forms of instructional media, including slide sets, expendable kits, and instructional kits, to DeKalb County schools. Among the subjects covered in the approximately forty sets of slides are insects, gemstones, meteorological instruments, and space exploration. The "Energy Center" at the Oregon Museum of Science and Industry offers a slide show for high school students that illustrates some uses of solar energy, especially in the Pacific Northwest.

Loan Materials

Science centers with collections sometimes loan artifacts and other materials to schools for instructional purposes. Occasionally, some centers also will make available scientific equipment and other materials for limited periods.

The Science Museum of Minnesota has about twenty-five teaching aid loan packets on protozoas, mollusks, rocks, butterflies, mammals, birds, fossils, and other fields. Reservations must be made at least one month before use, and loan materials may be kept only for one week. The museum delivers and picks up loans at St. Paul schools, but teachers outside of the city must make their own pickups and deliveries.

The Oregon Museum of Science and Industry operates a "Science Equipment Lending Library." Under the plan, schools may borrow scientific equipment, such as telescopes, a cloud chamber, lasers, and a spectrophotometer, from its research center.

Unfortunately, many museums have experienced difficulty with such loans, particularly with breakage and the failure to return objects promptly. As a result, some institutions require a fee and/or deposit; have strict requirements on usage; and restrict loans to a single day or two, with the pickup and return being the responsibilities of the schools.

Traveling exhibits are circulated in schools, communities, and regions by a number of science centers. Some are transported in vans and assembled for showing in schools, shopping centers, convention halls, and other such locations. Others are mobile vans with walk-through exhibits.

The Science Museum of Virginia in Richmond has operated a "Trans-science" mobile education unit throughout the state since 1973. The exhibit, which explores the potential of solar energy, is used primarily as a classroom supplement during the school year and for public display during the summer.

A "Traveling Museum" is circulated by the California Museum of Science and Industry. Three trucks carry space suits, slide shows, electronic materials, dry ice rockets, and other exhibit and demonstration material to distant parts of California. Seven lecture courses on such subjects as math, energy, minerals, space, and the human body are offered as part of the program. The traveling program is made possible by a grant from Union Oil Company.

Three of the most extensive traveling exhibit programs are offered by the American Museum of Science and Energy in Oak Ridge, Ontario Science Centre in Toronto, and Birla Industrial and Technological Museum in Calcutta. Funded by the U.S. Department of Energy, the Oak Ridge museum circulates two "Energy Encounters" exhibit trailers and several traveling exhibits on energy in the southeastern region of the United States. The Ontario Science Centre operates a traveling demonstraton and exhibit program for the Province of Ontario. In addition to school presentations, three exhibits— "Science Circus," "The Emett Things," "Mini-Circus"—are circulated even to remote regions of the province. The Calcutta museum has three mobile units—on energy, light and sight, and other subjects—that travel throughout West Bengal and other areas.

Science Demonstrations

An increasing number of science centers are giving science demonstrations and talks in classrooms, assembly halls, and other school facilities. Once offered free, most museums now charge a fee for such presentations because of increasing costs.

The Fernbank Science Center has a comprehensive list of traveling school programs, involving twenty-four elementary school, thirty-seven high school, and eight special education programs. All programs are presented without charge in schools that are part of the DeKalb County School System, since the science center is a division of the school system.

The New York Hall of Science has an outreach program for schools, presenting programs on earth science, general science, biology, and energy upon request and payment of a fee. In Seattle, the Pacific Science Center offers an in-school enrichment program that brings a teaching team, small participatory exhibits, lab lessons, and other resources to elementary schools for $175 per day. Up to ten classes per school may share in an all-day program.

"Suitcase OMSI" is an in-school demonstration program available through the Oregon Museum of Science and Industry. Using volunteers, the science center offers thirty-two presentations—carried in suitcases—in such fields as metric math, health, science, natural history, and energy. The basic fee is $12 per demonstration, with a minimum of two programs in one school per day.

A more elaborate form of school program is the "Traveling Science Show" provided by the Franklin Institute Science Museum. The science center can present eight demonstration programs in classrooms or assembly halls on such areas as chemistry, electricity, heat, and cryogenics. The charge for such shows ranges from $110 to $215, depending on the number of shows and demonstrations given and the number of schools served in a day. The Maryland Science Center in Baltimore has a similar program.

Enrichment Programs

A number of educational programs in which students "go to school" at science centers for a week have been initiated in cooperation with school systems. In some instances, the objective is integration as well as enrichment.

The Cranbrook Institute of Science offers a "Week at Cranbrook" program for sixth graders that includes scientific exploration at the museum, learning about astronomy, physics, anthropology, life sciences, and the earth. The project was conceived to give a broad understanding of science and to show the role of science in the everyday lives of young people.

Two innovative educational programs for elementary schools have been presented at the Franklin Institute Science Museum. The

"Paired Schools Science Enrichment Program," which began in 1967, was an intercultural joint venture with the Philadelphia School System that brought sixth grade classes from paired schools with varied racial and socioeconomic backgrounds to the museum one day a week for a six-week period. The youngsters spent the morning exploring such topics as electricity, light and color, action-reaction, water, air, and ecology with the aid of the museum's exhibits and demonstrations. The afternoon was devoted to a field trip to the waterworks or some other local resource related to the classwork. The second program, called "Project GOAL" for Greater Opportunities at Longstreth, brought the entire fifth and sixth grades from overcrowded Longstreth Elementary School to attend classes full time at the Franklin Institute for a school term. On one day a week, the sixth graders joined the school enrichment class.

The Museum of Science and Industry cooperated with the Chicago School System on an experimental "Academic Interest Center" program that involved a number of museums, subject areas, and facilities. Two classes of fifth graders spent four days together at the museum studying exhibits, seeing demonstrations, and receiving instruction from two school system teachers assigned to the museum. In addition to enriching learning, the program sought to further racial and social integration.

A somewhat different enrichment program is offered by the Maryland Science Center in Baltimore. Under the "Student Science Seminar Program," the education department presents Saturday lectures by prominent scientists and engineers at schools throughout the state. Approximately 1,000 high school students attend the eight-week programs.

In-Service Teacher Programs

More than half of the science and technology centers have in-service teacher programs. The programs are offered for two basic reasons: to acquaint teachers with the institution's science teaching resources and to provide science instruction to teachers. An in-service teacher institute or course may last from a half-day to a week or more. Sometimes, it carries credit from a local college or university.

Many in-service programs are presented in cooperation with the local school system. The one-day programs usually deal with more effective utilization of the museum and are offered free by the science center. A fee normally is charged for the longer programs with emphasis on science content. Among those institutions with credit-bearing teacher training programs are the Museum of Science in Boston, California Museum of Science and Industry, Lawrence Hall of Science, and Museum of Science and Industry.

Other Educational Programs

In addition to normal science activities in museums and special services to schools, most science and technology centers also are engaged in many other types of educational programs. They include preschool programs, independent study, internships, theater programs, science fairs and symposiums, nature trails, summer and wilderness camps, branch museums, curriculum development, special events, and radio and television programs.

Preschool Programs

Nearly every science center that offers courses also has some form of preschool program for young children. The degree of science content varies greatly. The Dallas Health and Science Museum has presented preschool courses, field trips, and events since 1957. The "Adventures in Science" program at Cranbrook Institute of Science seeks to introduce preschoolers from four to six to science through simple experiments and projects. The Fort Worth Museum of Science and History "Preschool" emphasizes the use of materials and equipment primarily in natural history, while the Lawrence Hall of Science's "Early Explorations" program focuses on biology. Omniplex in Oklahoma City relies heavily on exhibits, demonstrations, and simple experiments for its preschool and kindergarten programs.

Independent Study

About half of the science centers offer independent study. Fernbank Science Center offers facilities and guidance for independent exploration in seventeen areas, such as ecology, chemistry, biology, geology, and astronomy. The Pacific Science Center enables students to pursue special interests in independent study courses available in the evening, on weekends, and during the summer. In the "Science Career Program" at the St. Louis Museum of Science and Natural History, which began in 1959, research scientists are assigned to work with high school students on scientific investigations, the results of which are published.

The Oregon Museum of Science and Industry operates the "OMSI Research Center" for middle and high school students who are interested in conducting individual and group investigations and provides an equipped laboratory with advisors to assist about twenty-five students at any time on independent research projects in biology, physics, chemistry, optics, natural history, computer science, and other areas. A summary of the research projects is published each year. In Chicago, the Museum of Science and Industry has a "Science Club" that enables youngsters to use the museum's laboratory facilities for science fairs and other research.

Internships

Internship programs are being offered at an increasing number of science centers and, in most cases, are presented in conjunction with nearby colleges or universities and sometimes high schools. The programs usually relate to exhibit design, education programming, health education, curatorial work, and elementary science practice teaching. The "Explainer" program at the Exploratorium in San Francisco is a form of internship, although high school students are paid for interpreting exhibits to the visiting public.

Theater Programs

In addition to science demonstrations, an increasing number of science centers are using theatrical techniques, such as plays and puppet shows, to communicate scientific and technological information. In 1972–1973, the Museum of Science and Industry presented a "Science Playhouse" series of four plays with the assistance of professional actors from Goodman Theatre. A total of 28,743 people saw the fifty-six performances over a six-month period.

The Pacific Science Center offers puppet shows, one-person biographical characterizations (of Charles Darwin, Isaac Newton, Marie Curie), and a children's theater with plays on flight, volcanoes, animal trackers, and the metric system. The California Museum of Science and Industry has experimented with two theater groups: the "Space Place Players," which is a professional company, and the "Light Year Theatre Group," which is a community group. Both present entertaining plays on science topics, such as metrics, vitamins, and space travel. The Maryland Science Center presents special theater productions on such subjects as light, color, and locomotives.

Two traveling plays have been shown at a number of science centers. A one-man show entitled "Einstein—the Man" was produced by the American Museum of Science and Energy. Einstein was portrayed as a philosopher, humanist, and sensitive individual with a passion for violin music and a deep respect for theoretical and practical science. Another traveling science play was "Glass," developed with National Science Foundation support and performed by Otrabanda, a New Orleans-based theater troupe. The play, which received its title from the use of lenses to look into the past, sought to show the relevance and value of science to the average person.

Science Fairs and Symposiums

Science centers frequently are involved in organizing and presenting student science fairs, industrial arts exhibitions, Junior Achievement fairs, 4-H exhibits, and other such activities designed to display the achievements of young people. In many instances, the programs are presented in cooperation with local school systems or community groups.

The Museum of Science and Industry in Chicago is host to five such events each year: the Chicago Public Schools' Student Science Fair, Non-public School Science Exhibition, Exceptional Children's Week Program, Chicago Regional Industrial Education Exhibit, and Home Economics Fair. The California Museum of Science and Industry sponsors an annual Southern California Junior Science and Humanities Symposium, featuring papers by both outstanding students and scientists, and each spring, the Maryland Science Center presents a similar junior science and humanities symposium for nearly 210 leading eleventh-grade science students and teachers who are selected by local school systems throughout the state. The Oregon Museum of Science and Industry recently began a "Northwest Science Invitational" program for students in Oregon and Washington. The five-day science fair, which is cosponsored by the Oregon Science Teachers Association, is a successor to the regional

"Northwest Science Fair" held in the 1960s and disbanded in the early 1970s.

Nature Trails

Nearly every nature center and natural history museum has nature trails or gardens for conducted or self-guide tours. In some cases, the environmental areas are located adjacent to the science centers. Many institutions offer weekend tours to nearby parks, forests, and other areas having nature trails.

Fernbank Science Center of Atlanta has a 60-acre forest with one and a half miles of hard-surfaced trails as well as an experimental garden containing native and cultivated plants. The Charlotte Nature Museum, located in a 31-acre wooded park within the city limits, is planning to develop a 200-acre nature preserve and wilderness area ten miles from Charlotte. In addition to nature trails, weekend field trips and longer tours are offered as part of the environmental program. Other institutions with nature trail programs are the Cumberland Museum and Science Center in Nashville, Environmental Centers Inc. in West Hartford, Connecticut, Nature/Science Park in Winston-Salem, Science Museum of Minnesota in St. Paul, Children's Museum in Indianapolis, and Cranbrook Institute of Science in Bloomfield Hills.

Summer and Wilderness Camps

A few science centers have summer or wilderness camps operated as part of their educational programs. Other institutions sometimes will make use of camp facilities provided by members, companies, and other local organizations.

Cranbrook Institute of Science has a "Science Camp" program for boys and girls in the sixth, seventh, eighth, and ninth grades that operates for six weeks during the summer. Each camp period runs for one week and accommodates fifty campers. Each week is spent outdoors on the Cranbrook grounds and at the institute's two nature centers—one adjacent to the science center and the other a 33-acre nature study area along a nearby lake.

The Oregon Museum of Science and Industry offers a number of summer and wilderness camp programs. Founded as a paleontology camp in 1951, Camp Hancock in central Oregon now serves as a remote base for geology, botany, zoology, archaeology, astronomy, and other outdoor learning programs. OMSI also operates the Kiwanilong Camp, a wilderness camp on Long Lake near Warrenton, and the Optimist Acres Camp, a year-round nature facilities near Molalla.

Branch Museums

Relatively few satellite museums are operated by science and technology centers. In general, such "storefront" extensions have been expensive and not especially effective. It is difficult to present a visible minimuseum for neighborhood use with a few artifacts and exhibits, volunteer help, and a limited or irregular schedule.

The Oregon Museum of Science and Industry and the Housing Authority in Portland experimented with a "Drop-in Learning Center" concept in 1972–1973, but it met with limited success. The plan called for the operation of three inner-city centers staffed by college students who tutored in science and math on an individual basis.

The Rochester Museum and Science Center has received a better reception with its "Neighborhood Museum Project," which consists of six branch museums located in the inner city. Started as a staff project, the minimuseums have been reorganized under the Women's Council. Six volunteer teams bring the museum's resources to the neighborhoods.

A new approach to satellite museums is the "Museum on the Mall" concept started in 1978 by the Franklin Institute Science Museum in Philadelphia. The institute operates a small exhibit and live demonstration facility in a storefront space provided by the developer of the Gallery shopping mall. It has proved to be extremely popular. Joel Bloom, director of the Franklin Institute Science Museum, said the project is designed to reach a broader audience. "We see the mall as a teaser, a place to encourage people to come to the main museum."[20]

Curriculum Development

Although most museums are not involved in development of science curriculums or teaching tools for schools, the Lawrence Hall of Science, which is affiliated with the University of California at Berkeley, has specialized in such projects. In doing so, it has developed a large, specialized research staff and attracted numerous developmental grants from government agencies, foundations, and other sources. Its many science education research and development projects have dealt with elementary education for the disadvantaged, chemical education materials, outdoor biology programs, computer language techniques, science curriculum development, engineering instruction, and health activities. The "Discovery Van," a mobile workshop, takes the programs and materials developed at the center to surrounding schools for trial purposes.

Special Events

Many types of science-oriented special events involving science centers are held each year. In addition to science fairs and symposiums, they include scientific conferences, awards dinners, museum auctions, book fairs, and exhibit openings. Most such special events are educational in nature but frequently also have other purposes. The Oregon Museum of Science and Industry, for example, holds an annual "Aviation and Space Week" program that includes lectures, special exhibits, films, and even a paper airplane contest. In Seattle, the Pacific Science Center presents an annual "Science Circus" for a week or more during the Christmas season that involves special participatory exhibits, live science demonstrations, games, clowns, balloons, and other activities. The California Museum of Science and Industry honors the "California Scientist of the Year" and the "California Industrialist of the Year" at a gala banquet each year. The Franklin Insitute Science Museum has made awards for outstanding contributions to science and engineering since the parent institute was founded in 1824. Among those who have been honored on the annual "Medal Day" are Thomas A. Edison, Guglielmo Marconi, Niels Bohr, Glenn Seaborg, and Murray Gell-Mann. The museum also participates in an annual parkway festival and gives science demonstrations at various high-visibility events, such as Phillies baseball games.

A "Children's Science Book Fair," an "IR 100" exhibition and awards dinner, and a "Nobel Hall of Science" induction dinner are among the annual special events at the Museum of Science and Industry. Since initiated in Chicago, the science book fair has become a traveling event. The one hundred most significant new technical products developed during the year, as selected by *Industrial Research/Development* magazine, are honored at the "IR 100" program. The Nobel Hall dinner marks the addition of the latest

American Nobel Prize recipients in science in the museum's Nobel Hall of Science exhibit.

Radio and Television Programs

Science centers have recognized the tremendous potential of radio and television in reaching thousands and even millions of people, many of whom will never visit the institution. About half of the centers have media production facilities, and a much smaller number have broadcast studios. Amateur or "ham" radio stations can be found at science centers in New York, Philadelphia, Des Moines, Dallas, and other cities.

The Center of Science and Industry and American Museum of Science and Energy have the capability of making videotapes on new exhibits and programs available to local television stations. The Museum of Science and Industry circulates a weekly taped interview program to one hundred radio stations principally in the Great Lakes region and has a weekly children's science program on a local television station. Other science centers present courses, arrange for interviews, obtain public announcements, and take other steps to utilize radio and television for science education and publicity puposes.

School Group Visitations

One of the most difficult tasks at science and technology centers is handling arrangements for school group visitations. Since school groups constitute from a quarter to half of the total attendance at many institutions, the servicing of such groups has a direct relationship to the effectiveness of science centers.

"As every teacher knows," it has been pointed out, "taking the class on a field trip to a museum can be an enriching and enjoyable experience under the right circumstances. It also can be a nightmare if not planned properly."[21] Poor planning on the part of schools and/or science centers has contributed to unproductive class visits in many instances. The unannounced school bus that dumps children at the door and leaves them to wander aimlessly through the museum is the bane of science center directors and staff members. On the other hand, science centers can create troublesome headaches for teachers by not having tours, guides, demonstrations, programs, procedures, and/or facilities for handling school groups.

Too often, teachers are not aware of the exhibits, demonstrations, and program opportunities at science centers. They also may not be familiar with the procedures for school group visitations. Rather than blame the teachers, science centers should do everything possible to acquaint teachers with the offerings and requirements of their institution.

There are four parts to the school group visitation puzzle: procedures and facilities, advance preparation, the actual visit, and trip followup. A science center must do its share if it expects teachers and students to respond in an enlightened manner.

Procedures and Facilities

Every science center needs specific procedures and a number of facilities for the handling of visiting school groups. It should point out what the museum has to offer, how to register, how to reach the museum, and whether there are any special requirements. As for facilities, a science center should have a place for the teacher to check in; a checkroom, restroom, and lunchroom; bus parking; and possibly other facilities, such as orientation rooms and classrooms.

Program Information

The first step is to acquaint teachers and schools with the science center and its educational offerings. An institution must publicize its collections, exhibits, live demonstrations, theater programs, and other educational opportunities if it expects a teacher to take children out of class, arrange for a bus, involve parents, and make all the necessary arrangements for a museum visit. Many science centers do an excellent job in providing preliminary information through teacher's guides, program notices, and newsletters. Some also make school visits, use audiovisual presentations, and invite teachers to attend briefings, open houses, and in-service teacher programs.

One of the most common sources of information is the program mailer that describes the institution's offerings, procedures, and facilities for school groups. Among the effective publications in this respect are the "School Programs for Elementary and Secondary Schools" published by the Ontario Science Centre; "Guide to OMSI Happenings," Oregon Museum of Science and Industry; "School and Group Visits," Museum of Science, Boston; "A Guide for

Group Visits," Cranbrook Institute of Science; and "Teacher's Guide," Franklin Institute Science Museum.

To assist in scheduling and handling school groups, a science center should have a registration form with instructions on how to plan for a visit. In nearly all cases, booking arrangements also can be made over the phone. Most institutions require a school group reservation at least three weeks before the trip, although they frequently will accept reservations up to the day before a visit if space is available. Some science centers will not admit school groups that arrive without reservations, regardless of space considerations.

Typically a reservation form should include the name and address of the group, grade or age level, name of person in charge, number in group, date and time of arrival and departure, program choices, luncheon plans, and special needs. Alternate dates and program choices frequently are requested on a reservation form. Often a science center includes a price list and checklist for planning a visit to the center.

The Maryland Science Center tells teachers to call the reservation secretary but first make certain they have transportation to the center; at least one adult chaperone for every ten students; a definite idea of which program(s) they wish to attend; a specific date and time of arrival and departure (alternate dates may be necessary); full name, address, and telephone number of school and group leader; and grade, age level, and size of group.[22] The Baltimore center will not admit groups without reservations or without the required number of adult chaperones. Their typical school group tour lasts from two and a half to three hours and includes viewing the exhibits and attending exhibit hall demonstrations and one reserved program, such as the planetarium, theater, or classroom activity.[23]

Physical Facilities

In planning a school group visitation, it is helpful for a teacher to know whether the science center has a special entrance for school groups and checkroom, lunchroom, and parking facilities. The Ontario Science Centre has a special entrance and orientation area for school groups. The Museum of Science and Industry and American

Museum of Science and Energy also have special orientation facilities. The Rochester Museum and Science Center makes use of a multimedia show to interpret history, the natural sciences, and the museum's offerings.

Most science centers have school group lunchrooms with tables, vending machines, and sometimes hot food. The Franklin Institute Science Museum has a McDonald's fast-food restaurant in its building. In Chicago, the Museum of Science and Industry has a 600-seat school lunchroom as well as two public restaurants and an ice cream parlor.

Advance Preparation

A teacher should prepare for a museum visit by becoming familiar with the institution and its educational resources through reading, slide or films, or a previsit trip. Then the class should be briefed on the trip, with the teacher explaining some of the things to be seen and relating the information to classwork.

Previsit Literature and Trip

A trip to the science center can be most helpful in planning a school group visit, but it is not always possible. Therefore, most teachers rely upon published literature or previous contacts with the institution. A 1977 survey of museums in the plains and mountain states in the United States showed that 51 percent of the museums surveyed provided teacher's guides or other literature to schools; 17 percent made previsits to schools; 8 percent provided slides or filmstrips; and 4 percent furnished quiz sheets.[24] An Association of Science-Technology Centers study found that 15 percent of the responding science centers offered free literature on exhibits to teachers; 68 percent provided guidebooks; 13 percent supplied films; and 13 percent had quiz sheets.[25]

Class Preparation

It is highly desirable for a teacher to discuss plans for a museum visit with the class and to attempt to relate what will be seen to the classwork. Too often the school group visit has little or no connection with what is being studied at the time. A specific program for the museum visit should be planned and suggestions should be made on how to use free time. Some teachers will develop a series of questions or use standardized quiz sheets obtained from the museum to guide students through the exhibits and programs. Others will show slides, films, or filmstrips in advance of the trip for orientation purposes.

The Museum Visit

A school group visit to a science center averages about three hours and usually includes checking in, seeing exhibits, attending a program, having lunch, and buying a souvenir or two. In some cases, the visit includes an orientation and/or tour.

Check-in and Orientation

The first step in any visit is for the teacher to check on the program and lunch reservations. Nearly all science centers have a registration counter for this purpose. This usually is accompanied by the checking of clothes and the placing of lunches in a container for later use. If a science center has an orientation program, a guide or volunteer then will tell about the museum and the exhibits and programs that the group will see; otherwise, the class teacher may perform this function.

Museum Program

The museum program will vary with the choices made by the teacher. In general, it includes looking at the exhibits, floor demonstrations, and at least one scheduled program. In some science centers, it may be possible to arrange for a guided tour of the exhibits and demonstrations. More often a teacher must rely upon his or her own resources, such as leading the tour, using quiz sheets, or providing maps for self-guided tours. The scheduled programs normally are given in theaters, which can accommodate many school groups at one time.

Lunch and Shopping

In most instances, school group visits overlap with lunch; thus many science centers have lunchroom facilities. Because of the large number of visiting school groups, every group is assigned a lunch period. This schedule can become a bottleneck when school groups arrive late or stay too long. Every visit should allow some time for buying a postcard, booklet, or souvenir. This usually takes place just before the bus leaves, causing a last-minute rush at the museum store.

Trip Followup

Most field trips end when the bus returns to the school, but ideally a teacher should lead a discussion of what was seen and heard at the science center and how the experiences relate to subjects being studied in class. This gives the teacher an opportunity to focus on certain aspects of the visit and to show interrelationships and implications of particular exhibits, demonstrations, and programs. Some science centers provide discussion sheets or bibliographies for class use.

Another effective teaching technique is to require students to prepare reports on the museum trip or on subjects related to exhibits and programs at the science center. Such reports also could be the basis of a class discussion.

A field trip is followed by a test in some classes. More often, the quiz sheets used for self-guided tours of exhibits and programs become the examinations or the basis for discussions. Standardized quiz sheets are available from science centers for this purpose.

School group visits to science centers should be helpful to teachers and students. They can make teachers aware of new scientific and technological developments and information presentation techniques that can be applied in the classroom. For youngsters, a museum visit can open the door to self-discovery, independent study, and possible career opportunities. Unfortunately, such visitations seldom reap all the potential benefits because of inadequate advance preparation, lack of a specific visitation goal, poor supervision, insufficient followup, and/or sloppy handling by the science center.

18

Membership Activities

Philanthrophy and Marketing

Nearly every science and technology center either has a member-ship program or has considered starting one. Membership is viewed primarily as a vehicle for philanthrophy, but it also is a marketing tool. Although a membership program is difficult and costly to initiate and operate, it has the potential of greatly increasing interest, participation, and support of the institution from the community it serves.

The benefits and problems of a museum membership program have been debated for at least a half-century. It has been argued, for example, that such programs cost more than they produce or that servicing members is a headache; often that is the case. Some publicly operated museums do not feel a membership program is compatible with a free admission policy. Despite these arguments, membership programs and memberships continue to increase.

In 1935, I. T. Frary, membership and publicity secretary for the Cleveland Museum of Art, stated that "Memberships do not grow spontaneously; they must be built up painstakingly and maintained by constant care and cultivation."[1] That advice is still appropriate today. When well planned and executed, membership programs can be major sources of income and goodwill. But they can be a costly disaster when poorly conceived and implemented.

There are at least seven good reasons for having a membership program, according to Richard P. Trenbeth, development consultant and former director of development and membership at the Art

Institute of Chicago: it is a way of locating people who are interested enough in science centers to make a donation; it provides a source of fairly predictable annual income; it enlists goodwill ambassadors who advertise science centers by word of mouth; it offers a reason for asking for contributions from the membership several times a year; it can encourage people other than known contributors to become involved; it can break down communication barriers between members and museum administrators; and it is a way of recognizing regular donors of substantial amounts.[2]

Trenbeth believes people become members of museums because they have a psychological need for "belonging" to an organization; they seek specific benefits related to personal needs; they are lonely and welcome communications from a museum; they hope to acquire knowledge of culture that can help them solve personal problems or bring them social prestige; and/or they have strong emotional attachments to the institution.[3] It is essential for the management and membership staff of a science and technology center to understand and cultivate such reasons for membership interest. Although it nearly always is possible to develop a nucleus of members with built-in interests, it takes substantial effort to attract and keep a large number of marginal-interest prospects as members. An institution's success in building and cultivating both types of members usually determines whether a membership program justifies the investment of time, effort, and funds.

A number of science centers have been extremely effective with their membership programs, in particular the Museum of Science in Boston, which has 11,000 members. The program in Boston more than covers its costs, producing some $435,000 in membership dues, including $175,000 from 430 corporate members. More importantly, it also serves as the stimulus for most of the approximately $2 million contributions, capital gifts, bequests, and trusts. Other science centers with successful membership programs include the Science Museum of Minnesota, St. Paul, 12,000 members; Franklin Institute Science Museum, Philadelphia, 7,000 members; Cleveland Health Education Museum, 3,800 members; Center of Science and Industry, Columbus, 5,500 members; Maryland Science Center, Baltimore, 2,800 members; and Oregon Museum of Science and Industry, Portland, 4,500 members.

Some institutions, particularly those funded by governments, prefer to have "friends" organizations instead of membership programs. Such groups are primarily supportive in purpose and may be outside the museum's organizational structure. Friends usually do not elect the board of trustees nor do they receive all the benefits normally given to members, but they do require the same kind of diligent attention and cultivation for maximum returns.

Membership Operations

To function effectively, a membership program needs the specific and constant care of a membership office or department. The membership agency may operate as an independent department or as a branch of another department, such as the director's office, business department, or development department. Either approach is workable under the proper conditions.

Typically, the membership office is headed by a membership director, coordinator, or secretary. The office force usually includes a secretary and one or more clerks. Members and volunteers also may be involved in solicitations and mailings, but the office staff normally provides direction, keeps the records, and supervises the activities.

At some museums, the members elect the institution's governing board and officers. More often, the members have their own organization and officers. A number of science centers have membership programs consisting partially or entirely of "nonvoting" members, normally based on membership categories. In general, the officers of a museum membership program are concerned principally with member programming rather than the administration of the program.

The basic aims of any membership office should be threefold: to serve members, to promote interest in the museum and its activities, and to secure financial support for the institution. The functions of the office usually consist of prospecting, soliciting, servicing, renewing, fund raising, and recordkeeping.

Prospecting

An essential ingredient of any membership program is a prospect file or list. Such a file is a compilation of numerous lists obtained from a variety of sources. The basic prospect list should come from a science center's records of contributors, volunteers, friends, teachers, visitors, guests, and publication mailing lists. It should be supplemented by other lists furnished by museum staff members, trustees, and friends; compiled from organizational, city, and telephone directories; and supplied by other museums, community organizations, and mailing list houses. Although it sometimes is necessary to buy mailing lists, it frequently is possible to obtain membership lists from other organizations merely by asking for them or offering to exchange lists.

The names, addresses, and possibly other information, such as vocation, telephone numbers, family members, and notations on contacts made, should be kept on cards and/or in a computer file for prospects. This prospect file becomes the base for personal, telephone, and mail membership solicitations and also can be used for fund-raising purposes. Most prospect files are quite large to compensate for the relatively low return on most mailings.

Soliciting

The most effective membership solicitation is when one person asks another face-to-face to become a member of the museum. However, such soliciting is time-consuming and insufficient in most cases, except possibly in small communities. Therefore, the calling of prospects on the telephone by volunteers sometimes is used as part of a membership campaign. But whether the solictations are made in person or over the phone, it is vital that solicitors are oriented properly in their approach, in answering questions, and in handling contribution inquiries.

The most common form of membership solicitation is by mail; it also is the least effective from a success ratio standpoint. Yet, direct mail is a costly necessity. It is the only way to reach thousands of prospects, particularly those who are not known personally by membership workers. A favorable response of 1 percent is considered good, and anything higher is excellent. It often takes a mailing of 10,000 to produce around 100 members from a "cold" list; the percentage should be higher for lists of prospects who are familiar with the institution. Either way, followup mailings are essential.

A typical membership mailing includes a printed or autotyped letter of invitation from the board or membership chairman or the museum director; an explanatory leaflet or sheet describing the membership benefits and fees; a postage-paid business reply envelope; and a membership application, which sometimes is part of the leaflet or the envelope flap. A business reply envelope always produces a better response, and it is only necessary to pay the postage for envelopes that are returned. Whenever possible, membership mailings should be sent by bulk mail rates to minimize costs.

Most museums have a number of membership categories, and some institutions have several individual and/or corporate supporting memberships. Whatever the membership categories and dues, they should be clearly stated in check-off fashion on the membership application form. The form also should request payment by check with the application and indicate how the check should be made payable. An increasing practice is to allow member applicants to charge the dues against credit card accounts.

Any solicitation system also must have a means of acknowledgment. This usually consists of a printed note of welcome and a membership card (one for each qualified person in a family membership). The membership card should give the membership category and expiration date. Some museums use different colored cards for membership years or categories. For life and corporate members, some institutions send attractive certificates for framing.

Servicing

The nature and delivery of membership services affect both applications and renewals. As might be expected, they also influence annual giving, bequests, and other forms of contributions to the science center. Thus, the heart of any membership effort is the servicing of members.

A museum must be careful not to promise too much and deliver too little. Membership activities can have an adverse impact on the budget if they seriously erode the admission or sales income or if the social or publication costs are too high. On the other hand, a disappointed member is unlikely to renew his or her membership, contribute funds, or speak favorably of the institution.

A membership program is an institutional investment that requires sensitive leadership, a package of attractive activities, and fine tuning from time to time. The programming must appeal to a wide segment of the potential and current membership and adapt to changing conditions. A science center also must be careful not to cater to a few vocal members who can cause an unfortunate imbalance in the programming.

Renewing

The renewal of memberships should receive at least as much attention as the solicitation of member prospects. Too many museums assume that members will renew automatically because of their great interest in the institution and do not work at renewing memberships. Frequently they are surprised to see the rate of renewals dwindle. A science and technology center cannot take any member for granted. Even the most dedicated members will overlook renewing unless they are sent renewal notices and reminders. It also is necessary to reiterate the benefits of affiliation with the institution and to publicize membership events and services. Members must be convinced that membership is worth the investment of time and/or money.

Fund Raising

An effective membership program also can be the foundation for a museum's fund-raising activities. Members can be a devoted and supportive constituency that is responsive to annual giving programs, capital campaigns, and other requests for funds. Satisfied members also remember the institution in their bequests and trusts.

It takes more than member dues or contributions to support a science center. But such funds can be an important part of an institution's broader development program. Many museums conduct annual giving programs as part of the solicitation and renewal processes or during the course of the year in a manner similar to college and university alumni giving programs. Some also have specific tools to educate members on bequest needs and memorial opportunities. The Boston Museum of Science, for example, has published a number of pamphlets on how to make such contributions in wills, trusts, annuities, and even insurance policies.

Recordkeeping

The keeping of membership records ranges from rather simple files to complex computerized systems, depending upon the number of members, the extent of information on file, and the use of the records. A basic membership file might consist of a cash book, a file of membership cards arranged alphabetically, and a file of payment cards arranged by months or quarters. The payment cards are separated into "paid" and "unpaid" groupings, and those that are payable or delinquent are kept separate for billing and followup. At the beginning of each month, all payment cards for that month are transferred to the unpaid file. As dues payments are received, the cards are placed in the paid file. When using keypunching, similar information is stored on cards that can be categorized and separated as needed. In a computerized system, membership data can be stored on magnetic tapes or discs and be retrieved quickly for reporting, mailing, and other purposes. The use of computers, however, usually is considerably more expensive than a card file or keypunch system.

It generally is advisable to have a daily or weekly report as well as monthly and annual reports on new members, renewals, cancellations, and collections. Such information is extremely helpful in keeping abreast of totals and trends.

Some science centers also keep rather careful records of membership costs to determine the "true cost" of the program. In addition to direct expenses, they allocate overhead costs and even allow for a reduction in fee income resulting from free admission and sales discounts for members. When incorporating all such costs, however, an institution also should include contributions received through the membership program.

Membership Categories and Benefits

The categories and benefits of membership take many forms at science and technology centers. The dues structure usually is scaled, with more benefits going to those members paying higher fees.

Membership Categories

Nearly all museums with membership programs have at least three basic membership categories: student, individual, and family. Many also have one or more contributing memberships for a portion of the dues amounts to a donation. Museums also offer life and corporate memberships.

Student Members

Student memberships are for young people under 18 or 21 years of age. They are usually $5 or $10 annually and are designed to serve the educational needs of youngsters. Special activities, field trips, and summer programs frequently are developed for student members as a means of expanding their outlooks. Special clubs or groups, such as "Discoverers," "Explorers," or "Young Scientists," frequently are part of the program.

Individual Members

Individual adults who are 21 or older generally comprise this category. They normally pay $15 to $25 in dues and receive most membership benefits, including free admission, publications, discounts, invitations to open houses and exhibit previews, and access to the library and members' lounge.

Family Members

As family members, parents and their children under 21 share benefits similar to those for student and individual members. Some museums also have special activities for families, such as outings, films, and field trips. The annual membership cost ranges from $25 to $35.

Contributing Members

Contributing memberships usually start at $100 per year; a museum also may have supporting, sustaining, and governing memberships at $250, $500, $750, respectively. The purpose of these memberships is to provide support for the institution beyond the basic cost of membership. As the fee increases, so do the membership benefits.

Life Members

A life membership calls for a one-time payment of $500 to $1,000 (and sometimes more) and entitles the member to all the membership benefits for his or her life. An institution should be careful to set the life dues at a level that will cover servicing costs for a prolonged period but will not make such memberships prohibitive for most people. In general, life members are among the most loyal and supportive.

Corporate Members

Some science centers have one or more special membership categories for business and industry, which normally start at $100 and run to $1,000 or more. The benefits range from free admission, publications, and exhibit previews to access to the library, discounts on the use of facilities, and invitations to special showings and programs.

Membership Benefits

Membership benefits offered by science and technology centers generally include free admission, publication subscriptions, library and members' lounge privileges, member activities, exhibit previews, sales discounts, and/or special rates for educational and other programs. Some museums go beyond these basics.

Free Admission

The most common reason for becoming a museum member is free admission to the institution. Because of this appeal, a free museum sometimes has difficulty in starting a membership program. Other times, a museum may offer free admission to permanent exhibits or special exhibitions that normally require an admission fee.

Publication Subscriptions

Members nearly always are placed on the mailing list for the institution's newsletters, calendar of events, and/or professional journal. The newsletter and calendar are especially helpful in generating member interest in science center activities.

Lounge and Library Privileges

If an institution has a library or a members' lounge, access to such facilities usually is among the membership benefits.

Member Activities

Many science centers have special programs for members, such as tours, films, lectures, member nights, fashion shows, dinners, and exhibit previews. Behind-the-scenes tours of exhibits, collections, and work areas also may be part of member privileges. It is necessary to have a number of such activities to maintain and attract member interest.

Exhibit Previews

The opening of new exhibits always is one of the highlights of a membership program. Members enjoy being invited to exhibit previews and official openings, particularly for major permanent exhibits and traveling exhibitions. These events frequently involve luncheons, receptions, and dinners and sometimes the purchase of tickets.

Sales Discounts

A 10 percent or more discount on museum store purchases frequently is offered as a membership benefit. Such discounts generally cover souvenir and book sales but not food because of the small profit margin.

Special Rates

A membership program normally enables a member to register for educational classes and field trips and to attend lectures and film showings at reduced rates. The savings range from 10 to 25 percent.

Other Benefits

Some of the higher dues categories carry other benefits, such as book premiums, voting rights, and special showings. Corporate membership also may permit open houses for employees, shareholders, and customers; use of museum facilities for corporate board and shareholder meetings; and the funding of special educational exhibits.

19

Community Services

Responding to Community Needs

A science and technology center is more than a museum. It is a community resource with great potential for service beyond its walls. However, many institutions fail to engage in community activities that could be of great interest and value to the residents.

Community services present a number of policy, financial, and staffing considerations for any museum. It is not a simple decision, for example, to organize a science fair, to present minority programs, or to make facilities available to professional societies or hobby organizations.

Before plunging into community programming, a science center must consider several policy implications. How broad and deep should be the community involvement? What activities are compatible with the objectives of the institution? What should be the museum's position on activities outside the fields of science and technology?

Closely related to such policy questions is the matter of funding. How will community activities be financed? What impact will such services have on the funding of programs within the museum? When should a fee be charged for community programs or services?

Staffing of community activities also can present obstacles because of the outreach nature or evening scheduling of services. How can community programs receive necessary staff attention without increasing the museum staff? To what extent can the science center rely on volunteers from the community to supplement staff services? How can the necessary training for nonstaff members be provided effectively and economically?

These are questions that cause many science and technology centers to confine their community involvement to activities that are related to their science purposes; that are not costly; that can be handled by the existing staff with the help of volunteers; and that have direct benefits for the institution as well as the community. The decision to offer community services becomes much more difficult when the activities are not related to science; the costs are substantial; extra staff is needed; and/or the institutional benefits are marginal. But many museums make the investment of staff time and/or money because the services are needed by students, families, minorities, senior citizens, the handicapped, professional societies, hobbyists, and others in the community.

"In order for the museum to fulfill its role in the community, it must be a conscious part of the community and the community must be conscious that the museum is part of it," a Canadian museum management guide points out. "The public's cooperation and participation can only be assured on a continuing basis if the museum is actively aware of its community and the community is actively aware of the museum. This dual objective can be achieved through museum services and programs that are designed to meet the needs of its particular audiences."[1]

Many services are a response to identifiable needs in the community that benefit the institution very little. In other cases, the prime motivation frequently is self-interest—increasing attendance during a slow period; cultivating the interest of teachers and school administrators; increasing membership among students and parents; developing revenues from admission, registration, or rental fees; gaining acceptance among ethnic groups; obtaining favorable publicity for the institution; and increasing contributions and grants for programs and services.

Each institution must decide for itself on its course of action in community affairs. Justifiable reasons can be found for any level of community involvement. However, the trend appears to be toward more services in a wider range of fields that benefit both the community and the museum.

Types of Community Services

Community services offered by science and technology centers can be grouped into two broad categories: those that are science related and those that are not. Most institutions willingly offer the science-related variety of community services if the budget permits, but the same degree of interest does not exist for nonrelated community activities.

Science-Related Activities

Science centers frequently are involved in science-related services to schools; offer science courses, lectures, films, and tours; circulate science exhibits; stage special science events; operate speakers' bureaus; present radio and television programs; and offer museum facilities for use by scientific and technical societies, hobby groups, and industrial groups. Some of these community activities are part of the education department program, while others are handled by the exhibits or operations departments. Regardless of the organizational structure, community services deserve the attention of top management because of their importance in developing a rapport with the area served and meeting community needs.

School Services

Nearly all science centers provide some form of service to students, teachers, and schools. The services include science demonstrations, class instruction, publications, loan materials, suitcase exhibits, in-school programs, and teacher institutes. These services are discussed in chapter 17.

Courses, Lectures, Films, and Tours

Many types of science courses, lectures, films, tours, and other such programs are offered by science and technology centers. They are geared for children, adults, families, members, and/or other special groups. They frequently have an admission or registration fee to help pay the cost of such programs. Members usually are admitted free or receive a special rate.

Many institutions have special courses for preschoolers, elementary students, secondary students, and adults. Children's classes usually are on Saturdays and in the summer, and most adult courses are scheduled for weekday evenings. Among the subjects covered in courses at the Buhl Planetarium and Institute of Popular Science in Pittsburgh are astronomy, chemistry, life science, and model rocketry for children and astronomy, landscaping, and fishing rod construction for adults. In addition to its children's program, the Hall of Science of the City of New York has adult classes dealing with radio operation, invention, photography, and flying.

Although not as popular as they used to be, lectures on historical and current scientific and technological topics are presented at many science centers. One of the most successful programs at the Maryland Science Center in Baltimore is a travel lecture series, which regularly has capacity audiences. The Lawrence Hall of Science in Berkeley has a weekly lecture series, and the Museum of Science in Boston features periodic talks by leading scientists and engineers.

Most film programs are presented on Saturdays and Sundays. Some are aimed at children, while others are for adults and families. Sometimes the films are narrated by staff members or guests. The films range from children's cartoons to science documentaries. Science centers in San Diego, Detroit, and St. Paul have big-screen Imax or Omnimax theaters with day and evening showings that produce income for the support of the institutions.

Many types of field trips and tour packages can be found at science centers. The Oregon Museum of Science and Industry in Portland operates a wilderness summer camp; the Museum of Science and Industry in Chicago offers field trips to steel mills, auto assembly plants, scientific laboratories, and other such facilities, and the California Museum of Science and Industry has organized a tour to the Soviet Union for its members and friends.

Science Exhibits

Science exhibits developed by science and technology centers sometimes are shown in the community or circulated throughout a region. They range from portable tabletop exhibits to self-contained mobile vans. The Franklin Institute Science Museum in Philadelphia operates a minimuseum in a shopping mall, with exhibits and scientific demonstrations, the Science Museum of Virginia in Richmond takes an exhibit van around the state; the Ontario Science Centre in Toronto circulates exhibits throughout the province, and the American Museum of Science and Energy in Oak Ridge, Tennessee, is responsible for the operation of two walk-in vans and several traveling exhibits on energy that are shown in the southeastern region.

Special Events

A wide variety of special events involving science centers are held each year. They include conferences, science fairs, dinners, auctions, science plays, and exhibit openings. Some are staged at the museum, while others are presented elsewhere in the community.

Speakers' Bureau

Nearly every museum receives requests for speakers from schools, service clubs, women's groups, and other organizations. Whenever possible and appropriate such requests should be fulfilled. Because of the receptive audiences, some science centers actively solicit speaking engagements for key staff members to talk about the institution's activities or their professional fields. Such talks have both educational and public relations value. In operating a speakers' bureau, museums sometimes send out mailings to community organizations with suggested speakers and topics.

Radio and Television Programs

Science centers make extensive use of radio and television in reaching the community. Some museums have regular weekly programs on local commercial or public stations, while others make tapes or videotapes for circulating to a number of outlets in the region. The programs range from interviews to courses.

Use of Facilities

Many science centers cooperate with scientific and technological societies, hobby organizations, industrial groups, and others on the use of museum facilities for meetings, exhibits, conferences, and dinners. The Amateur Computer Society, Central Ohio Aquarium Society, Columbus Philatelic Club, and Columbus Amateur Radio Association have their meetings at the Center of Science and Industry in Columbus, Ohio. Among the temporary exhibits presented at the Oregon Museum of Science and Industry are the annual shows of the Oregon Agate and Mineral Society, Northwest Dairy Goat Association, and Portland Audubon Society.

Nonscience Activities

Most science and technology centers restrict their nonscience involvement either because such community activities are considered outside the institution's purpose or they are too costly. But almost all science museums do engage in some unrelated activities, such as community meetings, performing arts events, art exhibitions, holiday and folk festivals, and programs for minorities, senior citizens, and the handicapped. The justification is community need.

Community Meetings

The most common request is for the use of the institution's facilities—meeting rooms, the auditorium, dining rooms, and even office space. Most science centers are selective in allowing the free use or rental of such facilities and usually limit such nonscience usage to neighborhood groups, major community causes, and activities that cannot be accommodated elsewhere. A science center must be careful not to compete with commercial restaurants, hotels, and rental halls and thereby jeopardize its nonprofit tax status and antagonize local businesses.

Performing Arts Events

Music, drama, dance, motion picture, and other performances frequently are given at science centers as a community service. The Museum of Science and Industry presents a year-round series of free concerts by the Chicago Chamber Orchestra and offers a summer program of lagoon promenade jazz concerts in cooperation with local high schools and the City Department of Human Services. Modern dance programs, plays, and concerts are scheduled throughout the year at the Ontario Science Centre, and the Lawrence Hall of Science has weekend film showings for young children, teenagers, and adults.

Art Exhibitions

Art exhibitions at science centers take two forms: those that are science based and those that are more general in nature. Most science museums are receptive to science-oriented art exhibitions but are somewhat cool to other types of art shows. However, exhibitions of children's art, design, ethnic art, photography, senior citizens' art, and sculpture are becoming more common. In recent years, the Museum of Science and Industry has presented exhibitions of kinetic sculpture, neon art, posters, fiber art, laser sculpture, anamorphic art, photography, industrial design, prints, computer art, and children's, minority, and senior citizens' art.

Holiday and Folk Festivals

Some institutions are deeply involved in holiday and folk festivals, many of which occur during the summer or around Christmas. The Rochester Museum and Science Center annually has two programs of native crafts, dances, and choral performances: the "International Folk Festival" and "Holiday Folk Festivities." The Center of Science and Industry, Ontario Science Centre, and Museum of Science and Industry also have Christmas programs. "Christmas Around the World Festival" at the Museum of Science and Industry began in 1941 and involves nearly 40 ethnic groups that decorate sixteen-foot trees and present free holiday theater performances as well as 200 choral groups and a daily international buffet service.

Minority Programs

Special programs for blacks, the Spanish-speaking, native Americans, and other minorities are increasing among science centers. In most cases, they are not science based, although they occasionally include science elements. They usually are exhibitions and theater programs dealing with the artistic talent, cultural heritage, and achievements of specific minority groups. The Franklin Institute Science Museum and California Museum of Science and Industry have organized exhibits on black achievement; the Ontario Science Centre has presented a major Indian exhibition; and the Museum of Science and Industry annually sponsors a "Black Esthetics Festival" and a "Hispanic Festival of the Arts."

Senior Citizens Programs

As the aged population increases, many science centers are offering special rates and programs for senior citizens. Among the special programs for the elderly are social activities, evening open houses, special tours, conferences, films, and exhibitions of their art and crafts. A number of institutions have exhibitions similar to the annual "Golden Arts Fair" presented by the Museum of Science and Industry. The Rochester Museum and Science Center took a positive look at aging in a Sunday afternoon film, lecture, and discussion program entitled "Tomorrow is Not Far Away."

Handicapped Programs

Science centers also are taking steps to make their facilities and programs more accessible to the physically and mentally handicapped. They are installing ramps, elevators, and toilet facilities for visitors with wheelchairs; providing for those with visual, hearing, and other impairments in the design of exhibits; and presenting special exhibitions and programs for the handicapped. A number of institutions have art exhibits and theater programs featuring the handicapped during the annual "Exceptional Children's Week" in May.

Educational and Cultural Centers

Science and technology centers have become more than science museums in responding to community needs. In many instances they have been transformed into educational, cultural, and/or community centers.

It is only recently that "the idea has taken root that the museum has an obligation to the whole community, regardless of age, or type, or intellectual capacity," Molly Harrison pointed out in an International Council of Museums manual on museum organization. She said an "unenthusiastic public" is being won over by "a missionary spirit afoot in museums."[2]

The Spring Hill Center museum conference summary stated that "One of the educational purposes of museums is to provide extended opportunities for those out of school to expand their horizons and to provide a structured way of viewing the past, the present, and the implications of the future."[3] The conference report added, however, that "museums are too quick to criticize public schools" and are "very far behind the best schools in understanding the process of learning." Some conference participants argued that museums "are for all the people and could serve every single person in the community," while others responded that "they already operate at near-capacity attendance and that 85 percent of the audience is highly educated and prepared for the museum experience."[4] The report concluded that "It is probably the responsibility of the school system to do the preliminary education or cultivation of the interest in the artifacts and interpretation of culture. Then it can be a reasonable responsibility of the museum to enrich and expand the informed interest."[5]

Science centers and other types of museums are being swept up in the community movement, according to Richard L. Bunning, director of employee education for the Samaritan Health Service in Phoenix. "However, I fear that museums are often in much the same dilemma as those in the field of adult education: the most frequent participant is the individual with advanced levels of education, while the individual with less education, who needs it the most, is less likely to become involved."[6] Bunning urged museums to take steps to overcome this problem, such as projecting a concerned and involved community image through a greater public relations effort; offering classes, speakers, discussion groups, and other activities aimed at neglected segments of the community; sponsoring camera, stamp, antique, and other clubs related to the museum's theme; opening museum libraries to the public; organizing tours and traveling exhibits; and cooperating with other organizations on matters of community interest.[7]

It is obvious that science centers should not be cultural islands in their communities. They must venture beyond their property lines to carry science education into the community. Sometimes it also is necessary to cross the invisible dividing line between science and nonscience in reaching the disadvantaged, the elderly, the handicapped, the disinterested, and others who need the scientific, educational, and cultural services of a science and technology center.

20

Evaluation

The Evaluation Dilemma

Nearly everyone talks about the increasing need for evaluating museum exhibits and programs, but relatively few institutions have made a serious effort to measure the effectiveness of such activities, particularly from a learning standpoint. Many museums, including science and technology centers, argue that they simply cannot afford the time and money needed for systematic testing. They also have a lingering doubt that it is possible to measure the effectiveness of exhibits and programs or that the test results justify the time and cost.

Most exhibits and programs are planned intuitively and evaluative studies are rare. Watson M. Laetsch, former director of the Lawrence Hall of Science in Berkeley, believes it is not the cost or the evaluative techniques that inhibit systematic planning and testing but rather the attitudes of museum directors and staff members.[1] Frank Oppenheimer, director of the Exploratorium in San Francisco, differs sharply with such views on the practicality and effectiveness of evaluative studies in museums. He does not believe museum exhibits can be evaluated with traditional techniques. Most evaluative efforts use academic models that are not applicable to museum operations, he stated. The informal and subsurface learning that takes place in a museum and through exhibits differs considerably from that of the controlled classroom environment, according to Oppenheimer.[2]

Joel N. Bloom, director of the Franklin Institute Science Museum in Philadelphia, admits that evaluation of the museum experience has not been perfected but feels that science centers and other museums "must include meaningful evaluation procedures in our operating programs and engage in innovative research into the mechanisms of museum learning."[3]

All too often, evaluation degenerates into measuring what is easy and accessible rather than what a program is actually attempting to do. The primarily visual learning which takes place in the museum cannot be evaluated properly by simply copying and adapting techniques designed for the left-raised, linguistically based information transfer found in the classroom and lecture hall. Appropriate evaluation of the museum experience in its own terms is difficult. Furthermore, the translation and interpretation of the results of such studies into usable data for the museum professional is more difficult still. Yet, what other way is there to determine that the allocation of our limited resources is being done with maximal intelligence and integrity?

. . . The status of current research tools for investigating the cognitive and expressive aspects of museum learning and the interpretation of their results is still in its empirical and theoretical infancy. If we are committed to the idea that museums are not purely passive respositories of the culture of the past, but that museums are significant actors in the societal learning process, making important and singular statements necessary for a more complex understanding and intelligent decision-making, dare we continue to play blind man's bluff?[4]

C. G. Screven, professor of psychology at the University of Wisconsin-Milwaukee and a pioneer in the museum evaluation field, defines evaluation as "the systematic assessment of the value (worth) of a display, exhibit, gallery, film, brochure, or tour with respect to some educational goal for the purpose of making decisions (continue it, redo it, stop it, throw it out, avoid it in the future, etc.)."[5] Such evaluative studies are based on what he calls a "goal-referenced approach." The term "goal" refers to measurable learning or performance outcomes shown by museum visitors as the result of their exposure to a museum exhibit or program.

Screven has pointed out that museums are rather difficult places in which to seek a predictable educational impact:

Unlike public schools, public museums are informal learning environments where most "learners" are voluntary. In other words, visitors must be reached while moving freely through the halls. They are free to look at particular exhibits, but they are also free to ignore them, to misunderstand, to misread their diagrams, or whatever. Most visitors have little or no idea why the objects or paintings on display (except for well-publicized glamor items and some technology exhibits) are important. The typical museum audience varies widely in interests and backgrounds, has limited time, is often physically exhausted, and is frequently overwhelmed and confused by too many sensor inputs. Finally, it is difficult to control the order in which visitors view certain materials whose meaning depends on the order in which they are seen.[6]

Given these circumstances, is it reasonable to attempt serious communication of substantive ideas, concepts, and values to the general public? Screven believes the answer is "yes" if certain important prerequisites can be achieved. The museum director, designer, curator, educator, or planner must be willing to follow a few rules of the "educational game" that provide a basis for evaluation.

Screven has stated that at least three basic questions must be considered in evaluating the educational impact of exhibits: "What impact do you want? How will you attempt to achieve these objectives via your exhibit? How will you know if any of these methods or materials have had the desired impact on your intended audience?"[7] Screven, in presenting the first question, is concerned that the goals of most exhibit planners usually are general and must be broken down into specifics to be useful in evaluation. The second question is related to which exhibit content and techniques are most appropriate to achieve the goals. The third question, however, is the principal evaluation component. "What can the visitors do (report, describe, identify, compare, express, etc.) after exposure to the exhibit that they could or would not do beforehand?", Screven asks.[8]

Thus, the goal-referenced approach evaluates exhibits in terms of their intended goals and, if necessary, adjusts their design until the goals are attained. Screven said this evaluation is not possible unless the exhibit planner has clearly defined the educational or performance outcomes of the exhibit.

In 1968, the American Institutes for Research in Pittsburgh conducted a major study for the U.S. Office of Education on a traveling exhibit entitled "The Vision of Man" that dealt with the nature and accomplishments of many federal programs. The exhibit covered about 5,000 square feet and included over forty exhibit units on such topics as microwave energy, desalination, radio telescopy, nuclear energy, polymerization, and space exploration. The study evaluated the exhibit's effectiveness based on three types of variables: exhibit design variables (physical characteristics of the exhibit itself), such as the amount, readability, and legibility of verbal material, the use of audiovisual communication, total amount of time required to view exhibit materials, internal location and sequence of displays, and use of "constant" and "dynamic" models; exhibit viewer variables (audience characteristics and viewing conditions), such as the age/educational level, sex, socioeconomic level, and IQ of the audience, the amount of science background of the audience, initial attitude or level of interest of knowledge of the audience, and the viewing time and extent to which viewing is voluntary; and exhibit effectiveness variables (observations and tests of viewers), such as the ability to attract attention and hold interest, the ability to bring about a change in attitude, level of interest, or level of open-end concept knowledge (recall), open-end factual knowledge (recall), multiple-choice knowledge (recognition), or exhibit-specific knowledge. The first variables were independent or experimental, the second were control variables, and the third were dependent or criterion variables.[9]

One of the outcomes of the elaborate study was a so-called "Three Factor Theory of Exhibit Effectiveness." The theory reflected characteristics of exhibits that set them apart from other educational media, such as the voluntary nature of the audience. A comprehensive theory of exhibit effectiveness, according to the study, must concern itself with initially attracting viewers to the exhibit; maintaining their attraction throughout the exhibit; and maximizing the amount of relevant learning or "influence" that is achieved on the part of the viewer.[10] "If the exhibit is weak in any of these three areas, the chances of it achieving its stated objectives would appear to be greatly lessened," the report stated.[11] Thus, to be effective, an exhibit must exercise sufficient control over the viewers' behavior so that he or she is attracted, stays, and learns.

Types of Museum Testing

Museum testing ranges from simple visitor surveys to complex experimental studies, which seek to provide helpful information about the nature and preferences of visitors, the popularity and effectiveness of exhibits, and the educational impact of programs. The information may confirm what already is known or suspected or produce new data that may influence management decisions, exhibit designs, educational programming, and other aspects of a museum's operations.

Museum studies fall into two general categories: visitor research and experimental evaluation. Descriptive surveys of the visitor population are the most common, while evaluative studies are the most difficult and costly. Which is the most valuable depends upon the institutional need, while the most reliable is a reflection of the care in which a study is designed and implemented.

Visitor Research

Visitor surveys, sometimes called audience research, are conducted by many science and technology centers to gather descriptive data about their visitors. They usually are prepared and conducted by museum staff members on a minimum budget, although some extensive visitor surveys have been carried out on a contract basis.

Visitor Profiles

The primary aim of most visitor research is to obtain information about visitors, such as their age, sex, education, occupation, home town, income level, number in the party, why and how they came, and frequency of visits. Sometimes, an in-depth followup survey of selected subgroups of visitors is conducted by mail or phone.

Such demographic information is helpful to museum management in developing a profile of the population being served and determining how it compares with the institution's targeted market. Periodic visitor surveys also would show whether there is a seasonal change in the attendance and whether there are any marked differences in the visitor profile over a number of years. Such data can be useful in modifying and improving promotional efforts, exhibits, educational programs, and membership solicitations.

Visitor Feedback

A more imaginative and valuable visitor survey is one that obtains information about visitors' impressions, interests, and expectations. Such feedback about the museum's exhibits, courses, fees, demonstrations, employees, food services, films, parking, library, restrooms, and other aspects of its operations can be extremely revealing about the institution's exhibits, services, and facilities.

Before becoming complacent or rushing to correct apparent deficiencies, however, a science center must consider the source and method by which the data were gathered. When people are questioned in a survey, they frequently give answers that are expected and do not reveal their true feelings. In indicating which exhibits they liked the best or disliked the most, they often overlook the exhibits that had the greatest educational impact. And when asked to be critical, they may become gracious or condemning. Therefore, although visitor feedback is helpful, it is not always reliable and must be weighed with other factors in making decisions.

Experimental Evaluation

Evaluative studies are more scientific and formal than visitor surveys, but they also are less numerous and more difficult. They are based on controlled experiments that generally give them greater validity. However, in the free-flowing environment of museums, some control elements—such as the pretest and posttest—can influence the results.

Museum evaluations usually are conducted to improve an exhibit, program, or service. They normally include experiments to validate hypothetical relationships between components of the expected and actual outcomes. As pointed out in a Stanford University study, "Experimental control permits casual attribution by isolating the cause of observed differences between groups. However, practical considerations in evaluation often prohibit complete experimental control. Evaluators and museum personnel must determine the most critical extraneous variables and control them. Differences in uncontrolled variables should not present plausible rival explanations for observed differences between groups."[12]

Among the words of advice from the Stanford study are that

Resources must be invested wisely. Rather than sampling all cells in a factorial design, an evaluator should only examine conditions that are feasible and potentially acceptable to museum personnel. For example, to investigate the efficacy of active and passive exhibits, an evaluator might examine only static art exhibits, interactive art exhibits, and interactive science exhibits, if static science exhibits are *a priori* unacceptable. An evaluator should also be prepared to limit the investigation to selected subsets of the visitor population if all groups are not equally interesting or available.

Designs must be carefully selected. Experimental control must be weighed with generalizability and feasibility. The best evaluation rarely provides all the information necessary for policy decisions. Incompletely controlled experiments with limited generalizability are even less likely to be useful.

In experimental studies, experimental and control groups must be equivalent before exposure to the experimental treatment. Critical pre-test differences between the experimental and control groups make the interpretation of post-test differences equivocal. . . .

Problems associated with demographic differences between groups are avoided in within-subject designs. In within-subject designs, the same visitors are given a pre-test, exposed to the exhibit, and then given a post-test. A delayed post-test may also be administered to measure long-term retention. . . .

Another potential pitfall associated with collection pre- and post-test scores on the same visitors is that investigators may be tempted to compute gain scores as a measure of learning (pre-test minus post-test). . . .[13]

Since time and money are limited, it is inevitable that trade-offs will occur in evaluation design. A tightly controlled experiment to answer a few explicit questions may be too costly and/or too limited to form the basis of a generalization. On the other hand, more general survey techniques may be employed with interviews, questionnaires, and observations of visitors and on manipulation of components. Such data may not allow for casual inferences about the museum's impact, but the data may be easier to collect and just as useful to the museum staff.

In 1967, Screven divided evaluation into two types: formative and summative.[14]

Formative Evaluation

Many researchers believe that formative studies should be made before an exhibit is completed or a program is launched. This would enable museums, they argue, to use the results to change and improve elements of the exhibit or program to achieve its intended effects on visitor learning and performance. Actually, museums rarely perform formative evaluation of exhibits and programs. In general, they consider it an unnecessary waste of time and money.

But formative evaluation has some advantages, especially with the increasing pressure for accountability. In formative studies, low-cost mockups of proposed exhibits are shown to visitors to determine their effects on visitor attention, learning, and attitudes. Visitor input at this early stage can isolate difficulties and pinpoint the need for modifications, thereby making an exhibit more effective. Such a procedure was followed successfully by the Lawrence Hall of Science in developing its optics and astronomy exhibits.

Summative Evaluation

Most exhibit evaluation is summative in nature; it takes place after the exhibit is completed and operating. Summative studies seek to determine whether an exhibit is accomplishing what was intended. The results of such evaluation also can help decide whether an exhibit should be changed, continued, or removed. Even though it may be too late to alter the exhibit being evaluated, a summative study can be useful in planning other exhibits.

Screven outlined the stages in summative evaluation as defining the intended audience and the educational objectives of the exhibit; developing reliable measures of visitor knowledge and attitudes to reflect these objectives; pretesting visitors; observing visitor learning performance after exhibit exposure; analyzing and interpreting data; and adding or changing the exhibit under evaluation.[15] Theoretically, the procedure is sound. From a practical standpoint, however, many museum people question whether the measurements are reliable and whether it is possible to pretest visitors without influencing their behavior when they know they will be observed or post-tested.

Survey and Evaluative Techniques

A variety of techniques are available for visitor surveys and experimental evaluations. Some techniques are relatively simple and inexpensive, while others are quite complex and costly. However, more sophisticated and expensive testing does not necessarily produce better results.

In most instances, visitor surveys are based on questionnaires and interviews. Such surveys usually are conducted by the museum staff, are low cost, and provide helpful descriptive information about the museum's visitors.

Evaluative studies also make use of questionnaires and interviews, but generally in a more structured form and in a controlled environment. In addition, experimental investigations may employ other techniques, such as observation, pressure sensitive mats, and interactive devices. The objective is to obtain scientifically based information to improve an exhibit, program, or service. The experiments are designed to validate hypothetical relationships between the expected and actual outcome.

Questionnaires

A simple one- or two-page questionnaire that seeks demographic information about visitors is the most common survey technique. Visitors normally are handed such questionnaires after they have toured the museum and are about to leave. They are asked to complete and return the questionnaire before departing. To obtain a random sampling, every tenth, twentieth, or fiftieth visitor usually is asked to take part in the survey.

Such surveys most often are designed to obtain a profile of visitors to the museum or their opinions on exhibits, facilities, and services. Typical questions deal with the age, gender, education, occupation, hometown, and income level, as well as number in the party, why and how they came, frequency of visits, best liked and least enjoyed exhibits, employee attitudes, and other exhibits they would like to see at the museum. Such information can be extremely helpful in planning, promoting, and improving a museum's services.

At least three other types of questionnaires are used in evaluative studies: the completion questionnaire, open-ended questionnaire, and multiple-choice questionnaire. The first two are recall testing techniques, while the third is a recognition quiz.

The completion and open-ended questionnaires are designed to measure a visitor's ability to recall scientific concepts, principles, or other information covered in an exhibit. In a "concept" questionnaire, the objective is to determine whether exhibit viewers retained "the big picture" even though they may not be able to verbalize all the supporting details. A "knowledge" questionnaire, on the other hand, requires more specific recall of learned information.

Multiple-choice questionnaires consist of a series of questions with four or so alternative answers for each. Visitors are asked to select the correct answer from a list of choices. Recognition items generally are not as difficult as recall questions. Reading of the alternatives sometimes acts as a catalyst to prompt a correct response or to eliminate certain choices. The "exhibit-only" questionnaire is a form of mutliple-choice questionnaire, but it seeks to minimize prior knowledge and dispense with control groups and pretest correction by asking questions that can be answered only by recalling specific items from the exhibit.

Interviews

Both visitor surveys and evaluative studies make frequent use of interviews. Interviews may replace or supplement questionnaires and other techniques. In visitor surveys, followup telephone interviews sometimes are utilized on a selective basis to obtain information in greater depth. In evaluative projects, interviews are commonly used as the primary or secondary information-gathering technique.

Interviews can be structured or unstructured; make use of completion, open-ended, or multiple-choice questions; and be conducted in person or over the telephone. Unlike questionnaires, interviews enable researchers to explore meaningful areas uncovered during the course of an interview.

The most productive interviews generally are those that have specific questions but allow for discussion of related issues that emerge during the interviews. Such interviews require skilled interviewers and can be time-consuming and costly. Quite often the interviews are taped for later playback and analysis.

Observation

The observance of visitor behavior in museums has been an effective research tool for more than half a century. One of the first museum studies based on observation was conducted at the now-defunct New York Museum of Science and Industry in the mid-1930s under the supervision of Arthur W. Melton. Without questionnaires or interviews, he sought to determine the most effective types of exhibits for a science and technology museum by observing visitors as they moved through the museum's exhibit halls and recording what caught their attention and how much time they spent examining exhibits and reading labels. His early studies showed the importance of movement and participation in attracting visitor interest.[16]

Observation of museum visitors usually takes place in three ways: from a hidden location behind an exhibit, through closed-circuit television, or by following visitors unobtrusively and taking notes of their actions. Such trackings are extremely helpful in determining the pulling power of various exhibits and techniques, but they do not measure educational effectiveness. It simply is not possible to find out through observation how much a visitor has learned from an exhibit.

Pressure Sensitive Mats

Because museum visitors normally do not act naturally when they know they are being observed, a technique was developed to record their footsteps remotely. Sometimes known as the "hodometer" system, the use of pressure sensitive mats can show the traffic pattern in an exhibit hall.[17] Although rarely employed, the floor grid measurement technique can be useful in recording the number and location of footsteps in a room.

The system has two parts: a network of pressure sensitive electrical switch mats that are laid across the floor and a bank of counters in a cabinet behind the scenes. The mats, which are covered by a carpet, are activated when someone steps on them. The counters record the pressure points to indicate which exhibits have the greatest appeal. Because the system is automatic, it does not require an observer or operator. But the system is costly to install and provides limited data.

Interactive devices have become important tools in the evaluation of participatory exhibits. Basically, they fall into two broad categories: added devices that compel the visitor to respond actively to questions and exhibits that involve the visitor in investigating scientific principles and technological applications through direct manipulation of the apparatus. They sometimes are used in combination with questionnaires, interviews, and/or observation techniques.

C. G. Screven pioneered the use of numerous gamelike devices in his studies at the Milwaukee Public Museum. One of the various interactive response devices he has added to exhibits is a portable electronic punchboard. A museum visitor answers programmed questions by using a stylus to punch holes at appropriate places on paper answer sheets placed on the punchboard. A cassette tape player—used individually by each visitor and plugged into the punchboard—guides the visitor's attention to relevant exhibit features and asks questions to be answered on the punchboard. The tape stops automatically and restarts only after a question is answered correctly. Visitor performance is measured by counting answer holes in the sheets.

Gamecards are a simpler version of the punchboard technique. Commercially available multiple-choice answer cards are used with the cassette tape. In this case, the visitor must restart the tape after each question and answer. It is possible to give the technique a more gamelike quality and immediate feedback by using a "magic pen" which prints one color for correct answers and another color for incorrect responses.

Another interactive device is the free-standing self-testing machine. Using colored slides, a free-standing machine presents various shapes, words, and pictures that can be matched to other shapes, words, and pictures by selecting one of four buttons. The buttons turn green for correct and red for incorrect choices. The questions may reflect factual knowledge, conceptual understanding, or interests of the visitor. A tape recorder registers each visitor's responses, how long he or she remains, error frequencies, etc. The device may be placed anywhere in the museum to monitor visitors prior to, during, or following exhibit exposures; can make use of bells, counters, and other techniques to attract visitors to interact;

and can be activated at a central location with coded tokens distributed at specific exhibits. When using tokens, a sequence of questions is selected automatically based on the exhibit from which the tokens were obtained.[18]

A slightly different type of portable testing machine was used in an evaluative study conducted by Minda Borun at the Franklin Institute Science Museum in Philadelphia. A booth resembling a study carrel that contained a rear-screen projected slide test and a push-button response mechanism was used to present a cognitive-affective test to a random sample of museum visitors. Test questions appeared on slides of museum exhibits, and visitors were asked to enter their responses by pushing buttons on a multiple-choice response device that punched the answers on cards. The machine was designed to make the test more like a game than a schoolroom experience.[19]

Representative Surveys and Studies

An increasing number of visitor surveys and evaluative studies are being conducted by science and technology centers despite the reservations of some museum professionals. In general, the research has been helpful in obtaining a better understanding of museum visitors and the effectiveness of exhibits. The following representative visitor surveys and evaluative studies and their findings demonstrate the value and difficulties of such research.

Visitor Surveys

Nearly every science center has conducted some form of visitor survey to gather information about its visitors and their feelings about the exhibits, facilities, and/or services. The approaches differ, but the objective is the same: to better understand, attract, and serve the visiting public.

Unfortunately, many visitor surveys are virtually worthless. In 1973, Ross J. Loomis, associate professor of psychology at Colorado State University and a museum researcher, wrote an article, titled "Please! Not Another Visitor Survey," in which he stated:

The title above depicts the way a number of museum personnel feel about visitor surveys. In fact, it is virtually a direct quotation from a museum director who was confronted with the idea of doing some audience research. The director assumed that visitor research meant a "survey," and for him, and many others in the museum field, this form of research has some credibility problems. There is a pervasive belief in our culture that institutions show their accountability, in part, to the public by "surveying" them, no matter how ludicrous the results.[20]

Museum of Science and Industry, Chicago

Three or four visitor surveys are conducted each year to develop visitor profiles during different seasons and to sample their impressions of the museum's offerings. Approximately 1,000 visitors—every twentieth person—are asked to complete questionnaires before leaving the building. Sometimes the interview technique is substituted with visitors being asked the same questions from the questionnaire. The results are virtually identical.

The visitor surveys have revealed that about half of the museum's four-million visitors come from the Chicago metropolitan area and half from elsewhere, including every state and about one hundred countries. The surveys also have shown almost an even split between children and adults and between male and female visitors. The length of a typical visit is about three and one half hours, more than triple the time normally spent on a museum visit. About 75 percent of the visitors have been to the museum before.

Because of the heavy tourist attendance in the summer, the Museum of Science and Industry also conducts a license plate check every August to determine the number of different states represented in the parking lot at any one time. The count consistently is around thirty-five states. This information is useful in documenting the museum's high out-of-state attendance during the vacation season.

Ontario Science Centre, Toronto

Visitor surveys also are conducted periodically by the Ontario Science Centre. The first such survey, conducted in the fall of 1975, was designed by the Ontario Ministry of Industry and Tourism. It was a fourteen-question interview survey used to obtain a profile of some 500 visitors. The questionnaire had two parts: a section completed by the interviewer (indicating the time of day, date, day of the week, sex of the person interviewed, and the interviewer's name) and the questions answered by the visitors (number of visits, length of visit, transportation, how center was brought to visitors' attention, main reason for visit, rating of facilities and services, general impression of center, ability to see everything, plans for future visits, number in party, age of visitors in party, hometown, other sights seen in Toronto, and length of stay in Toronto).

About 70 percent of the visitors were at the Ontario Science Centre for the first time. Of those who had been there before, 15 percent were showing the center to relatives or friends from out of town. Approximately 12 percent of the visitors said that the visit was planned as a treat for their children. Almost half of the visitors said they heard about or visited the center because of favorable comments from friends, while about 20 percent made the visit after reading about the center in a magazine, newspaper, or brochure.[21]

Oregon Museum of Science and Industry, Portland

A different type of visitor survey was carried out by Shabtay Levy, director of exhibits at the Oregon Museum of Science and Industry in 1976. The objectives were threefold: to determine the age and sex of visitors attracted to particular exhibits, the length of time spent at each exhibit, and the average viewing time for the exhibits. Twenty-six diverse exhibits were selected for the survey. Volunteers were stationed near the exhibits over a three-month period (March, April, and May). More than 17,000 visitors were observed in 202 survey reports.

The survey disclosed that the largest portion of visitors came from the elementary school group of 6 to 14 years of age (43 percent), followed by adults 26 to 60 years (28 percent), high school students 14 to 18 years (13 percent), college students 18 to 26 years (7 percent), preschoolers (7 percent), and seniors 60 or more years (2 percent). Of the total, 52.8 percent were male and 47.2 percent female.

The study indicated that 27.5 percent of the visitors saw all twenty-six that were included in the exhibits survey and spent an average of one minute and fourteen seconds in front of each exhibit. The popularity-time (P.T.) factor—which is the multiplication of the percentage of people who visit an exhibit by the time they spend there—averaged 32.4 for all the surveyed exhibits. In general, the participatory exhibits were most popular, with visitor participation ratios ranging from 35 to 98 percent.[22]

Evoluon, Eindhoven

A survey of paying visitors was undertaken by A. H. Boerdijk, director of the Evoluon in The Netherlands, to examine attendance fluctuations in yearly, seasonal, monthly, weekly, and daily totals. Using reports from 1967 through 1974, he found that the science center's attendance figures were constant; more than half of the visitors came during the April-August season; the monthly totals did not change significantly from year to year; the peak month was July and the low month was January; the weekly figures fluctuated with holidays and feast days; and Saturdays and Sundays drew more visitors most of the year, but weekdays surpassed weekends during the summers.[23]

The survey revealed that the duration of a visit went down as the attendance increased, largely because the center became more crowded and it was more difficult to see and interact with the exhibits. The turning point appeared to be when more than 1,500 visitors were in the building.

Two other surveys were performed in conjunction with Boerdijk's study. In 1973, every fiftieth visitor was asked to complete a profile questionnaire. A total of 7,562 visitors were surveyed. The study showed that only 6.3 percent of the visitors were from the Eindhoven area; most visitors came with their families (64.3 percent); it was the first visit for 67.8 percent; students were the largest single category (50.3 percent); the most common educational level was secondary school (40.1 percent); and the top age group in attendance was 12 through 19 years of age (43.9 percent).

The Philips organization that operates the Evoluon also hired the Dutch Institute of Public Opinion and Market Research to do a followup study involving 600 of the visitors who completed the questionnaires. The survey sought visitor opinions about the science center. Results showed that 92 percent believed the Evoluon has succeeded in demonstrating the role of science and technology in society and culture; 36 percent replied that the exhibits were "interesting and educative" and 34 percent said they were "worth looking at"; 14 percent indicated the exhibits were "too technical" or "difficult to understand"; and 66 percent have advised others to visit the Evoluon as a result of their experience.[24]

Evaluative Studies

Experimental studies are more difficult, time-consuming, and costly than visitor surveys, but they also can be more valuable if properly conducted. Such evaluative investigations normally seek to validate beliefs about an exhibit, program, or service. The studies may take place during the developmental stage or once the exhibit, program, or service is offered to the public. They are especially helpful instruments in improving museum exhibits.

New York Museum of Science and Industry

The first experimental study at a science and technology center was conducted in the mid-1930s at the New York Museum of Science and Industry under the supervision of Arthur W. Melton. It was a pioneering study that sought to determine the most effective types of exhibits for such museums. The museum died from lack of support, but Melton's work lives on in museum circles.

Using the observation technique, Melton found that movement and participation had great attention-getting impact on exhibit viewers. He learned, for example, that a massive piece of machinery—a gear-shaper—would receive much greater attention when operated. Objects near the gear-shaper also benefited considerably, while those farthest away received less attention. His study findings showed that although the average time spent by visitors in the exhibit hall increased from 258 to 278 seconds when the gear-shaper was put into operation, the overall increase was not commensurate with the increased attention received by the gear-shaper portion of the exhibit. Melton concluded that the moving gear-shaper had distracted attention from other exhibit units in the room and that the result was not necessarily a net educational gain.

The installation of a pushbutton-operated new exhibit unit—a pinion gear—drew considerable attention and caused a redistribution of the traffic pattern in the gallery. Melton said such attention-getting exhibit units can have a detrimental effect on other objects in the room, but when "a particularly attractive exhibit draws more visitors into a certain section, this distraction-effect may be more than counter-balanced by the increased attendance."[25]

Melton said that participation techniques stimulated interest in exhibits of both children and adults. "The cranks were not just gadgets to turn; they heightened interest to such an extent that more label reading was done."[26]

Melton concluded his study with two pleas, based on his experiments. First, the behavior of visitors toward all the objects in an exhibit hall must be studied whenever a change is made in an exhibit unit. It should be possible to increase interest in one aspect of an exhibit without impairing interest in other sections. The goal should be to increase the total exhibit interest. Second, exhibits should not be made attractive at random. By making some relatively unimportant portion appealing, it is possible to seriously damage the educational service of a truly significant aspect of an exhibit.[27]

Museum of Science, Boston

Two sociologists from Brandeis University—Robert W. Weiss and Serge Boutourline, Jr.—in 1963 used observation and interview techniques to determine how people traveled through the museum in Boston. They found that no one in the sample came to the museum alone; people apparently like to share museum-going with others. The study also showed that museum visitors did not attempt to see every exhibit, and that they seemed "to want to develop a sense of the [exhibit] hall as a whole, rather than a sense of each individual exhibit." However, when there was something particularly attractive or prominent in an exhibit hall, visitors tended to examine it. When exhibits presented materials similar to those already seen, visitors usually moved along unless the units were directly in their path.

Weiss and Boutourline observed that visitors generally took a fairly straight route through an exhibit hall, rarely crossing from side to side unless forced to do so by a major exhibit unit in the hall. As reported by Arthur W. Melton in 1935 on traffic in an art museum, the researchers noticed that visitors tended to walk through a hall along its right wall in a counterclockwise direction, even though exhibits may be designed for left to right viewing.

The most attractive exhibits from a visitor standpoint, they found, had a common characteristic: they could be seen only in the museum. The least interesting were those exhibits with written and illustrative materials that could be found in a book without coming to the museum. To learn which exhibits were remembered, Weiss and Boutourline showed photographs of the exhibits to fifty-five visitors as they left the main hall of the Boston museum. The exhibit units that were most prominent—a Mercury space capsule replica and a lighthouse beacon lens—were cited by fifty-four of the fifty-five respondents. Murals, which were positioned at a number of places along the walls, were remembered by only about 50 percent of the people. Some exhibits, such as a two-dimensional panel exhibit on elements, were recalled by 20 percent or fewer of the visitors.

Interviews with museum visitors revealed that children and adults, as well as men and women, respond differently to the same exhibits. Different types of exhibits often are evaluated on different bases, according to the study. A model of a yacht, for example, was praised for its beauty, while the lighthouse lens was appreciated because it was something that normally is not available for examination. Some exhibits, such as the "atomic pile," caused visitors to feel that they should try to learn more about them. "It seems to us," Weiss and Boutourline stated, "that exhibits which attempt to communicate complex or difficult ideas are often responded to with respect, but without comprehension."[28]

Lawrence Hall of Science, Berkeley

The Lawrence Hall of Science has been the scene of a number of evaluative studies using visitor-operated demonstration machines and open-ended exploratory activity booths as well as questionnaires and interviews.

In a 1973–1974 study by Laurie P. Eason and Marcia C. Linn of the Lawrence Hall staff, an attempt was made to determine whether two exhibit types—machines and booths—were effective in communicating information about optical principles; whether one exhibit type was more effective than the other; and how effective was a presentation combining the two exhibit types. "Exhibit effectiveness" was defined as "a measurable transmission of information about scientific principles from the exhibits to the visitors."[29]

The machines were self-contained, operated in a standing position, and had pushbuttons and knobs to change the position of a mirror, lens, or prism with respect to a light source. Backlighted text and simple diagrams further explained the principles presented. In the booth format, the visitor was seated and each booth was equipped with a light source and experimental objects, such as mirrors, lenses, prisms, protractors, and rulers, attached to a platform by long chains. The visitor manipulated the equipment directly to observe the particular optical principle and to explore further aspects of the phenomenon. Brief instructions, diagrams of the principle, and leading questions were provided by backlighted panels.

Two evaluation measures, a short multiple-choice questionnaire and a five-minute apparatus-oriented interview, were used to determine which exhibit technique was more effective in conveying information about optics. The test population consisted of 740 fifth through eighth graders visiting the Lawrence Hall in organized school groups. The youngsters were divided into experimental and control groups for comparative purposes. Measurable transmission of information was determined to occur when students in the experimental group correctly responded to significantly more questions than those in the control group.

The study found that students exposed to the exhibits did better than the control group, but that the difference between machine and booth approaches was not significant. Students who experienced the combined presentation of machine and booth exhibits had slightly higher scores than those who were exposed to only one of the exhibit types; however, analysis of variance revealed no significant difference between the two groups.[30]

Franklin Institute Science Museum, Philadelphia

A one-year pilot study was conducted over 1975–1976 by researcher Minda Borun at the Franklin Institute Science Museum to develop models for testing visitor response, provide usable information to the museum staff, and test the feasibility of a large-scale investigation of science museums. The ultimate stated goal was to incorporate regular channels of visitor feedback into museum practice "so that the museum becomes a flexible, self-evaluating, self-correcting institution in touch with the needs and desires of its public."[31]

The study began with the administration of a goal-rating scale to define "effectiveness" in terms of the goals of the Franklin Institute museum staff and visitors and of other member institutions of the Association of Science-Technology Centers. The resulting list of goals served as a guide in regulating objectives for testing visitor response. Units of measurement and measuring techniques were developed to provide the baseline data, which served as an indication of the museum's "effectiveness" at the beginning of the study.

Five questionnaires were administered to samples of from 100 to 200 visitors, soliciting information on the motivation for the visit, visitor interests, exhibit attendance, exhibit preferences, and building orientation. A detailed mail-back questionnaire also was given to a sample of teachers of visiting school groups.

Attitudinal change and information transfer were measured by means of a multiple-choice test. The affective section sought to determine interest in science, understanding of the impact of science and technology in daily life, attitude toward science and technology, and attitude toward the museum. The cognitive section was designed to measure learning of basic science information through a test of vocabulary, experience, and concept formation.

Visitors were given a quiz in a portable testing machine, which consisted of a small booth equipped with a rear-projection screen and pushbutton response mechanism. Test questions appeared on slides of museum exhibits, and visitors pushed buttons to indicate their choices on punchcards.

The baseline data served as a basis of comparison for measuring the impact of possible changes and improvements. Learning and attitude measures were based on a comparison of previsit and postvisit test scores. A random sample of 250 visitors received the pretest and another sample of 250 visitors were posttested, making a total baseline sample of 500 visitors.

The study results showed a clear increase in knowledge from the pretest to the posttest scores but a decrease in positive feelings about the museum, science, scientists, and technology. The latter was attributed largely to bicentennial-related construction during the testing period that disappointed many visitors. The investigation also revealed that building orientation was a major problem for visitors.

The study also indicated that visitors preferred the more complex exhibit hall, having thirty to forty displays per room, as opposed to halls with a sparser, more contemporary display style; there was a negative correlation between popularity and number of background colors; participatory exhibits were more popular than other exhibits; there was a strong inverse correlation between instructional power and number of participatory devices; and pushbuttons, when separated from other participatory devices, accounted for most of the negative correlation with instructional power. It was concluded that pushbuttons, which worked well to attract attention, were appropriate in introductory and transitional exhibit areas but were not effective aids to the communication of scientific facts and principles.

The report concluded with an assessment of the museum's effectiveness in achieving its stated goals. The museum was found to be highly effective in terms of teaching basic science concepts but was much less effective with respect to the affective goals of stimulating and developing curiosity, interest in science, and positive attitudes toward science and technology. The last result was attributed, in part, to the extremely high initial scores of entering visitors.

Borun cited the need for more experimental studies in museums:

Clearly, we need systematic studies of visitor response to controlled variation of exhibit components in order to understand how color, lighting, labelling, placement of objects, number, size, and complexity of objects, and type of display can attract and hold visitors' attention and contribute to measurable "cognitive gains."

It's important for us to come to understand the nature of the visitors' reaction to a museum visit to explore the instructional effectiveness of different types of exhibits, to appreciate the dynamics of visual and interactive learning, and to incorporate this understanding into the exhibit-making process.

In addition, we need comparative studies of data collected in a range of institutions, in order to distinguish general principles of exhibitry and visitor response from the effects of specific museum contexts. The pilot study discussed here and the other references cited are a beginning; there is much yet to be done.[32]

Shabtay Levy of the Oregon Museum of Science and Industry expressed somewhat the same feelings in saying his institution's visitor survey documented what had been "a matter of speculation." "It is a step further in the ever-increasing importance of evaluating exhibits and programs in the science center of today," he added. "Evaluation is necessary if we want to improve professionally and give better service to our visitors. Systematic evaluation always involves subjective judgments, but these judgments are least support by a data base."[33]

Although doubts persist about the reliability, value, and cost of evaluative studies, there appears to be little doubt that museums, including science and technology centers, will become more involved in such studies in the future. The need to justify the expenditure of limited funds and the desire for more effective techniques will require better evaluation before and after the development of museum exhibits and programs.

Appendix

Representative Science and Technology Centers

Alfa Cultural Center, Apartado Postal 1177, Monterrey, N. L., Mexico

American Museum of Science and Energy, Oak Ridge Associated Universities, P.O. Box 117, Oak Ridge, Tennessee 37830

Birla Industrial and Technological Museum, 19A Gurusaday Road, Calcutta 700 019, India

Buhl Planetarium and Institute of Popular Science, Allegheny Square, Pittsburgh, Pennsylvania 15212

California Museum of Science and Industry, 700 State Drive, Los Angeles, California 90037

Center of Science and Industry, 280 East Broad Street, Columbus, Ohio 43215

Charlotte Nature Museum/Discovery Place, 1658 Sterling Road, Charlotte, North Carolina 28209

Cleveland Health Education Museum, 8911 Euclid Avenue, Cleveland, Ohio 44016

Cranbrook Institute of Science, 500 Lone Pine Road, Bloomfield Hills, Michigan 48013

Cumberland Museum and Science Center, 800 Ridley Avenue, Nashville, Tennessee 37203

Dallas Health and Science Museum, Box 26407, Fair Park, Dallas, Texas 75226

Des Moines Center of Science and Industry, Greenwood Park, Des Moines, Iowa 50312

Detroit Science Center, 52 East Forest Avenue, Detroit, Michigan 48201

Deutsches Museum, D-8000 Munich 26, West Germany

Dutch Institute for Industry and Technology, Rozengracht 224, 1016 SZ Amsterdam, The Netherlands

Evoluon, Noord Brabantlaan 1a, 5652 LA Eindhoven, The Netherlands

Exploratorium, 3601 Lyon Street, San Francisco, California 94123

Fernbank Science Center, 156 Heaton Park Drive, N.E., Atlanta, Georgia 30307

Rueben H. Fleet Space Theater and Science Center, 1875 El Prado, San Diego, California 92103

Franklin Institute Science Museum, 20th Street and Benjamin Franklin Parkway, Philadelphia, Pennsylvania 19103

Don Harrington Discovery Center, 1200 Streit Drive, Amarillo, Texas 79106

Impression 5 Museum, 1400 Keystone, Lansing, Michigan 48910

Korean Children's Center, Yook Young Foundation, San 3039, Neung-Dong, Seongdong-Ku, Seoul, Korea

Lawrence Hall of Science, University of California, Centennial Drive, Berkeley, California 94720

Maryland Science Center, 601 Light Street, Baltimore, Maryland 21230

Mid-America Center, 400 Mid-America Boulevard, Hot Springs, Arkansas 71901

Museum of Science, Teodoro Roviralta 55, Barcelona 22, Spain

Museum of Science, Science Park, Boston, Massachusetts 02114

Museum of Science, 3280 South Miami Avenue, Miami, Florida 33129

Museum of Science and Industry, 57th Street and Lake Shore Drive, Chicago, Illinois 60637

Nagoya Municipal Science Museum, 17-22 Sakae 2-Chome, Naka-Ku, Nagoya 460, Japan

National Museum of Science and Technology, Lahore-31, Pakistan

National Science Museum, 2 Warydong-Dong Chongno-Ku, Seoul 110, Korea

National Taiwan Science Hall, 41 Nan-Hai Road, Taipei, Taiwan, Republic of China

Nehru Science Centre, Dr. E. Moses Road, Worli, Bombay 400 018, India

New York Hall of Science, P.O. Box 1032, Flushing Meadows-Corona Park, Flushing, New York 11352

North Carolina Museum of Life and Science, 433 Murray Avenue, Durham, North Carolina 27704

Omniplex, 2100 N.E. 52nd Street, Oklahoma City, Oklahoma 73111

Ontario Science Centre, 770 Don Mills Road, Don Mills, Ontario M3C 1T3, Canada

Oregon Museum of Science and Industry, 4015 Southwest Canyon Road, Portland, Oregon 97221

Pacific Science Center, 200 Second Avenue, North, Seattle, Washington 98109

Palais de la Découverte, Avenue Franklin D. Roosevelt, 75008 Paris, France

Science Museum, 928 Sukhumvit Road, Bangkok 11, Thailand

Science Museum, Exhibition Road, South Kensington, London SW7 2DD, England

Science Museum, Japan Science Foundation, 2-1 Kitanomaru Koen Chiyoda-Ku, Tokyo 102, Japan

Science Museum of Minnesota, 30 East 10th Street, St. Paul, Minnesota 55101

Science Museum of Virginia, 2500 Broad Street Station, Richmond, Virginia 23220

Singapore Science Centre, Science Centre Road, Singapore 22, Republic of Singapore

Technological Museum, Federal Commission of Electricity, Apartado Postal 18-816, Mexico City 18, D. F., Mexico

Visvesbaraya Industrial and Technological Museum, Kasturba Road, Bangalore-1, India

John Young Museum and Planetarium, 810 East Rollins Street, Orlando, Florida 32803

Traveling Exhibition Services

American Federation of Arts, 41 East 65th Street, New York, New York 10021

American Institute of Graphic Arts, 1059 Third Avenue, New York, New York 10021

Association of Science-Technology Centers Traveling Exhibition Service, 1016 Sixteenth Street, N.W., Washington, D.C. 20036

Collage Inc., 3228 Military Road, N.W., Washington, D.C. 20015

Gallery Association of New York State, Box 345, Hamilton, New York 13346

Independent Curators Inc., 1740 N Street, N.W., Washington, D.C. 20036

International Exhibitions Foundation, 1729 H Street, N.W., Washington, D.C. 20006

International Museum of Photography at George Eastman House, 900 East Avenue, Rochester, New York 14607

Museum of Modern Art National Program of Circulating Exhibitions, 11 West 53rd Street, New York, New York 10019

Oak Ridge Associated Universities, Museum Division, Traveling Programs, P.O. Box 117, Oak Ridge, Tennessee 37830

Pratt Graphics Center, 160 Lexington Avenue, New York, New York 10016

Smithsonian Institution Traveling Exhibition Service, A&I 2170, Smithsonian Institution, Washington, D.C. 20560

Visual Studies Workshop Inc., 4 Elton Street, Rochester, New York 14607

Western Association of Art Museums, Department of Circulating Exhibitions, Mills College, P.O. Box 9989, Oakland, California 94613

Notes

Chapter 1

1. Lee Kimche, "Science Centers: A Potential for Learning," *Science* 199 (1978): 270–273.

2. *Hands-on Museums: Partners in Learning* (New York: Educational Facilities Laboratory, 1975), p. 5.

3. Anne C. Roark, "The Playground of the Museum World," *The Chronicle of Higher Education* 17 (1979): 8.

4. Kenneth Hudson, *Museums for the 1980s: A Survey of World Trends* (New York: Holmes and Meier Publishers, 1977), p. 3.

5. Alvin Schwartz, *Museum: The Story of America's Treasure Houses* (New York: E. P. Dutton & Co., 1967), p. 222.

6. John Whitman, "More than Buttons, Buzzers and Bells," *Museum News* 57 (1978): 43.

7. Michael Templeton, "More than Buttons, Buzzers and Bells," *Museum News* 57 (1978): 44.

8. National Research Center of the Arts Inc., *Museums USA* (Washington, D.C.: National Endowment for the Arts, 1974), pp. 48–49.

9. National Center for Education Statistics, *1979 Museum Universe Survey* (Washington, D.C.: Institute of Museum Services, 1980).

10. Bernard S. Finn, "The New Technical Museums," *Museum News* 43 (1964): 24–25.

11. W. T. O'Dea and L. A. West, "Editorial," *Museum* 20 (1967): 150.

12. Frank Oppenheimer, "A Rationale for a Science Museum," *Curator* 11 (1968): 206.

13. *Evoluon* (Eindhoven: Evoluon, 1966), p. 2.

14. Loren D. McKinley, in *The Franklin Institute News,* Summer 1973, p. 17.

15. *Bylaws,* Association of Science-Technology Centers, Washington, D.C., 1973, p. 1.

16. Ad Hoc Committee on ASTC Membership, "Statement of ASTC Membership Criteria," in Final Report, Association of Science-Technology Centers, Washington, D.C., 1978, p. 1.

17. *Museum Accreditation: A Report to the Profession* (Washington, D.C.: American Association of Museums, 1970), p. 6.

18. H. J. Swinney, ed., *Professional Standards for Museum Accreditation* (Washington, D.C.: American Association of Museums, 1978), p. 12.

19. Francis Henry Taylor, "Museums in a Changing World," *The Atlantic Monthly* 164 (1939): 792.

20. Duncan F. Cameron, "The Museum, a Temple or the Forum," *Curator* 14 (1971): 11.

21. Ibid., p. 12.

22. George Basalla, "Museums and Technological Utopianism," *Curator* 17 (1974): 106–107.

23. Ibid., p. 107–108.

24. Edward P. Alexander, *Museums in Motion* (Nashville, Tenn.: American Association of State and Local History, 1979), pp. 74–75.

25. Howard Learner, "Science Museums: Educational or Promotional?", *Nutrition Action* 6 (1979): 9.

26. Ibid., p. 3.

27. Howard Learner, *White Paper on Science Museums* (Washington, D.C.: Center for Science in the Public Interest, 1979), pp. 51–52.

Chapter 2

1. Eugene S. Ferguson, "Technical Museums and International Exhibitions," *Technology and Culture* 6 (1965): 30.

2. Ibid., p. 31.

3. Ibid.

4. Ibid.

5. Gottfried Wilhelm Leibniz, "An Odd Thought Concerning a New Sort of Exhibition," in *Leibniz Selections,* Philip Wiener, ed. (New York: Charles Scribner's Sons, 1971), pp. 585–594.

6. Ibid.

7. Ibid.

8. Ferguson, "Technical Museums and International Exhibitions," pp. 31–32.

9. Ibid., p. 32.

10. Ibid., p. 33.

11. M. Y. Bernard, chairman, Conservatoire National des Arts et Métiers, letter, Oct. 10, 1978.

12. Ferguson, "Technical Museums and International Exhibitions," pp. 33–34.

13. Ibid.

14. Ibid., pp. 35–39.

15. Ibid.

16. Ibid.

17. F. Greenaway, *A Short History of the Science Museum* (London, 1959), p. 5; and Bernard S. Finn, "The Science Museum Today," *Technology and Culture* 6 (1965): 195.

18. George Basalla, "Museums and Technological Utopianism," *Curator* 17 (1974): 106.

19. Paul H. Oehser, *The Smithsonian Institution* (New York: Praeger Publishers, 1970), p. 87.

20. *Illustrated Guide Through the Collections,* Deutsches Museum, Munich, 1971, pp. 12–13.

21. Oskar von Miller, *Technical Museums as Centers of Popular Education,* Deutsches Museum, Munich, 1929, p. 4.

22. *Illustrated Guide,* pp. 12–13.

23. Von Miller, *Technical Museums as Centers of Popular Education,* p. 4.

24. Frantisek Jilek and Jiri Majer, *National Technical Museum,* National Technical Museum, Prague, 1967, pp. 1–5.

25. *Guiding Through the Collections,* Technological Museum of Industry, Crafts, and Trades, Vienna, 1972, pp. 1–2.

26. Ibid.

27. Richards, *The Industrial Museum,* p. 48.

28. *Greenfield Village and the Henry Ford Museum* (New York: Crown Publishers Inc., 1972), p. 9.

29. Ibid., pp. 98–99.

30. Ibid.

31. "Dr. Jewett Heads Science Museum," *New York Times,* Oct. 19, 1935.

32. Charles T. Gwynne, *Museums of the New Age,* Association for the Establishment and Maintenance in the City of New York of Museums of the Peaceful Arts, 1927, p. 11.

33. Ibid., p. 20.

34. Ibid., p. 23.

35. Ibid., p. 15.

36. "Scientific Items Gleam in New Home," *New York Times,* Feb. 11, 1936.

37. *The Museum of Motion,* New York Museum of Science and Industry, New York, 1936, p. 2.

38. Herman Kogan, *A Continuing Marvel; The Story of the Museum of Science and Industry* (Garden City, N.Y.: Doubleday & Co., 1973), p. 10.

39. Ibid., pp. 10–11.

40. Waldemar Kaempffert, *From Cave-Man to Engineer,* Museum of Science and Industry, Chicago, 1933, pp. 11–12.

41. Thomas Coulson, *The Franklin Institute from 1824 to 1949,* Franklin Institute, Philadelphia, 1949, p. 22.

42. Ibid., p. 31.

43. Ibid., pp. 30–31.

44. Robert P. Multhauf, "European Science Museums," *Science* 128 (1958): 514.

45. "A Visit to the Palais de la Découverte," *Revenue Du Palais de la Découverte,* July 1975, p. 37.

46. Ibid.

47. Richard O. Howe, "Planning for the Future," address to ASTC workshop, Indianapolis, Oct. 31, 1978, p. 1.

48. Ibid.

49. Viola L. Oberson, *Oregon Museum of Science and Industry* (Portland, Ore.: Metropolitan Printing Co., 1963), pp. 3–5.

50. Douglas P. Huegli, "Museum Miracle in Oregon," *Museum News* 38 (1959): 30.

51. Frank Oppenheimer, "The Role of Science Museums," in *Museums and Education,* Eric Larrabee, ed. (Washington, D.C.: Smithsonian Institution Press, 1968), p. 167.

52. Ibid., pp. 169–170.

53. Frank Oppenheimer, in Barbara Gamow, *Evolution of a Palace,* Palace of Arts and Science, San Francisco, 1970, p. 5.

54. Frank Oppenheimer, *Statement of Broad Purposes,* Palace of Arts and Science, San Francisco, Sept. 27, 1971.

55. *Fernbank Science Center,* Fernbank Science Center, Atlanta, Georgia, p. 3.

56. *A Brief Guide to the Evoluon,* Evoluon, Eindhoven, The Netherlands, pp. 2–3.

57. Ibid., p. 12.

58. *Facts About the Ontario Science Centre,* Ontario Science Centre, Toronto, p. 2.

59. J. T. Wilson, "Ontario Science Centre Attracts the Millions," *Canadian Geographical Journal,* March/April 1976, p. 5.

60. *Facts About the Ontario Science Centre,* p. 2.

61. *Your Guide to Japan Science Foundation,* Japan Science Foundation, Tokyo, 1964, p. 3.

62. *Science Museum Visitor Guide,* Science Museum, Tokyo, 1974, p. 1.

63. Amalendu Bose, "Objectives of a Science Museum," *Science and Culture* 44 (May 1978): 194.

64. Ibid.

65. Toh Chin Chye, in *Singapore Science Centre,* Singapore Science Centre, Singapore, 1972, p. 3.

66. R. S. Bhathal, "Science Museums Can Be Fun," *New Scientist* 65 (1975): 82.

Chapter 3

1. *Recommendations on a Museum of Science and Technology for Hong Kong,* Association of Science-Technology Centers, Washington, D.C., 1978, p. 11.

Chapter 4

1. Kenneth Hudson, *Museums for the 1980s: A Survey of World Trends* (New York: Holmes & Meier Publishers, 1977), p. 80.

2. Ibid., pp. 80–81.

Chapter 5

1. Jacques Barzun, "Museum Piece, 1967," *Museum News* 46 (1969): 17.

2. Ger van Wengen, "The Museum: Aims, Methods, and Evaluation," *Museums' Annual* 6 (1974): 27.

3. June Batten Arey, *The Purpose, Financing and Governance of Museums,* Spring Hill Center, Wayzata, Minn., 1978, pp. 3–4.

4. *Final Report,* International Meeting on the Planning of Museums of Science and Technology in Developing Countries, Manila, Dec. 11–15, 1978, p. 2.

5. Ibid., p. 1.

6. Theo Stillger, "Objectives of a Museum of Science and Technology," address to the First International Conference of Science and Technology Museums, Philadelphia, 1978, p. 1.

7. *Annual Report 76/77,* Center of Science and Industry, Columbus, 1978, p. 1.

8. *Policy Guidelines,* Science Museum of Minnesota, St. Paul, Oct. 20, 1978, p. 1.

9. *Objectives of the Museum of Science and Industry,* Museum of Science and Industry, Chicago, 1981, p. 1.

10. Amalendu Bose, "Objectives of a Science Museum," *Science and Culture* 44 (May 1978): 194.

11. Ibid.

12. Ibid.

13. Ibid.

14. *Maryland Academy of Sciences—Maryland Science Center,* Maryland Academy of Sciences, Baltimore, 1977, p. 3.

15. *Japan Science Foundation,* Japan Science Foundation, Tokyo, 1964, p. 3.

16. Ibid.

17. *Nagoya Municipal Science Museum Ordinance,* City Ordinance No. 27, Aug. 10, 1962.

Chapter 6

1. Helmuth J. Naumer, *Of Mutual Respect and Other Things* (Washington, D.C.: American Association of Museums, 1977), p. 10.

2. Francis Parkman and E. Laurence Springer, *The Independent School Trustee Handbook* (Boston: National Association of Independent Schools, 1974), p. 4.

3. Naumer, *Of Mutual Respect and Other Things,* p. 16.

4. June Batten Arey, *The Purpose, Financing and Governance of Museums* (Wayzata, Minn.: Spring Hill Center, 1968), p. 15.

5. Ibid.

6. John Henry Merryman, "Are Museum Trustees and the Law Out of Step?" *Art News* 74 (1975): 24.

7. Ibid., pp. 24–25.

Chapter 7

1. Daniel Raverso and Mavis Bryant, "How to Start a Museum" in *Thoughts on Museum Planning,* Reference Series 3 (Austin: Texas Historical Commission, 1978), p. 17.

2. Carl E. Guthe, *So You Want a Good Museum* (Washington, D.C.: American Association of Museums, 1973; reprint of 1957 edition), p. 14.

3. Ibid., p. 13.

Chapter 8

1. Frederick J. Turk, "Some Management Tips for Arts Organizations," *World* 12 (1978): 31.

2. Peter F. Drucker, *The Practice of Management* (New York: Harper & Brothers, 1954), pp. 7–17.

3. Ibid.

4. Ibid., pp. 341–343.

5. Ibid., pp. 343–346.

6. *Museum Management Tools,* final report to National Endowment for the Arts, Management Analysis Center Inc., Washington, D.C., 1978, pp. 1–6.

7. Ibid., p. 8.

8. Ibid.

9. Ibid., pp. 18–19.

10. Ibid., p. 118.

11. George MacBeath and S. James Gooding, eds., *Basic Museum Management* (Ottawa: Canadian Museums Association, 1975; reprint of 1969 ed.), p. 35.

12. H. Lawrence Wilsey, "Management Techniques" in *The Arts Management Handbook,* Alvin E. Reiss, ed. (New York: Law-Arts Publishers Inc., 1974), p. 291.

13. Ibid., pp. 291–293.

14. Ibid., p. 293.

15. Malvern J. Gross, Jr., *Financial and Accounting Guide for Nonprofit Organizations* (New York: Ronald Press Co., 1974), p. 309.

16. Ibid., p. 314.

Chapter 9

1. June Batten Arey, *The Purpose, Financing and Governance of Museums* (Wayzata, Minn.: Spring Hill Center, 1978), p. 10.

2. Daniel Traverso and Mavis Bryant, "How to Start a Museum" in *Thoughts on Museum Planning,* Reference Series 3 (Austin: Texas Historical Commission, 1978), p. 8.

3. *The Impact of Chicago's Museum Visitors on the City's Economy* (Chicago: Leon Burnett Co., 1977), p. 4.

4. *An Introduction to the Economics of Philadelphia's Cultural Organizations* (Philadelphia: Greater Philadelphia Cultural Alliance, 1975), p. 8.

5. *Minnesota: State of the Arts* (St. Paul: Governor's Commission on the Arts, 1977), p. 230.

6. G. Ellis Burcaw, *Introduction to Museum Work* (Nashville: American Association for State and Local History, 1975), p. 41.

7. Traverso and Bryant, "How to Start a Museum," p. 8.

8. Harold J. Seymour, *Designs for Fund-Raising* (New York: McGraw-Hill Book Co., 1966), p. 50.

9. Ibid., p. 54.

Chapter 10

1. Victor J. Danilov, ed., *Museum Accounting Guidelines* (Washington, D.C.: Association of Science-Technology Centers, 1976).

2. Robert N. Anthony, *Financial Accounting in Nonbusiness Organizations* (Stamford, Conn.; Financial Accounting Standards Board, 1978); and Accounting Standards Executive Committee, *Proposed Statement of Position on Accounting Principles and Reporting Practices for Nonprofit Organizations Not Covered by Existing AICPA Audit Guides* (Exposure Draft), American Institute of Certified Public Accountants, New York, April 1, 1978.

3. William H. Daughtrey, Jr., and Malvern J. Gross, Jr., *Museum Accounting Handbook* (Washington, D.C.: American Association of Museums, 1978), p. 93.

4. *Accounting Principles and Reporting Practices for Certain Nonprofit Organizations,* Price Waterhouse & Co., New York, 1978, p. 9.

5. Danilov, ed., *Museum Accounting Guidelines,* p. 8.

6. Malvern J. Gross, Jr., *Financial and Accounting Guide for Nonprofit Organizations* (New York: Ronald Press Co., 1974 ed.), p. 7.

7. Paul Bennett, *Up Your Accountability* (Washington, D.C.: Taft Products Inc., 1973), p. 60.

Chapter 11

1. Paul Goldberger, "What Should a Museum Building Be?" *Art News* 74 (1975): 33.

2. G. Ellis Burcaw, *Introduction to Museum Work* (Nashville: American Association for State and Local History, 1975), p. 146.

3. Daniel Traverso, "Do's and Don'ts for Museum Planning Groups," in *Thoughts on Museum Planning,* Reference Series 3 (Austin: Texas Historical Commission, 1978), p. 4.

Chapter 12

1. Robert G. Tillotson and Diana D. Menkes, eds., *Museum Security* (Paris: International Council of Museums, 1977), p. 2.

2. Douglas A. Allan, "The Staff," in *The Organization of Museums* (Paris: UNESCO, 1967; reprint of 1960 ed.), p. 63.

3. G. Ellis Burcaw, *Introduction to Museum Work* (Nashville: American Association for State and Local History, 1975), p. 100.

4. "Museums Fight Rising Toll of Inside Theft," *U.S. News & World Report* 84 (1978): 74.

5. "Art Institute Tells Theft of 13th Century Chinese Scroll," *Chicago Sun-Times,* Jan. 4, 1979.

6. Tillotson and Menkes, *Museum Security,* p. 10.

7. Ibid.

8. Denis B. Alsford, *An Approach to Museum Security* (Ottawa: Canadian Museums Association, 1978), p. 2.

9. Jack Leo, "A Basic Security Checklist for the Small Museums," *Texas Assn. of Museums Newsletter,* 1978, p. 8.

10. Tillotson and Menkes, *Museum Security,* p. 56.

Chapter 13

1. George MacBeath and S. James Gooding, eds., *Basic Museum Management* (Ottawa: Canadian Museums Association, 1975; reprint of 1969 ed.), p. 75.

2. Ibid.

3. Daniel Traverso and Marvis Bryant, "How to Start A Museum," in *Thoughts on Museum Planning,* Reference Series 3 (Austin: Texas Historical Commission, 1978), p. 17.

Chapter 14

1. *Webster's Seventh New Collegiate Dictionary* (Springfield, Mass.: G. & C. Merriam Co., 1972), p. 558.

2. Carl E. Guthe, *So You Want a Good Museum* (Washington, D.C.: American Association of Museums, 1973; reprint of 1957 ed.), p. 1.

3. George MacBeath and S. James Gooding, *Basic Museum Management* (Ottawa: Canadian Museums Association, 1975; reprint of 1969 ed.), p. 43.

4. Duncan F. Cameron, "The Museum, a Temple or the Forum," *Curator* 14 (1971): 12.

5. R. S. Miles, "The Public's Right to Know," *Nature* 275 (1978): 682.

6. B. Halstead, "Whither the Natural History Museum?", *Nature* 275 (1978): 683.

7. Guthe, *So You Want a Good Museum,* p. 2.

8. Dorothy H. Dudley and Irma Bezold, *Museum Registration Methods* (Washington, D.C.: American Association of Museums, 1958), p. 2.

9. MacBeath and Gooding, *Basic Museum Management,* p. 50.

10. Robert G. Chenhall, *Museum Cataloging in the Computer Age* (Nashville: American Association for State and Local History, 1975), p. 10.

11. G. Ellis Burcaw, *Introduction to Museum Work* (Nashville: American Association for State and Local History, 1975), p. 97.

12. MacBeath and Gooding, *Basic Museum Management,* p. 59.

Chapter 15

1. Theo Richmond, "Museum in the Front Line," *The Illustrated London News* 266 (1978): 56–57.

2. G. Browne Goode, *The Museums of the Future,* 1897 Annual Report, Part II, U.S. National Museum, 1901, pp. 243–262.

3. William K. Gregory, "The Museum of Things Versus the Museum of Ideas," *Science* 83 (1936): 585–588.

4. G. Ellis Burcaw, *Introduction to Museum Work* (Nashville: American Association for State and Local History, 1975), p. 116.

5. Carl E. Guthe, *So You Want a Good Museum* (Washington, D.C.: American Association of Museums, 1973; reprint of 1957 ed.), p. 28.

6. Burcaw, *Introduction to Museum Work,* pp. 117–118.

7. Robert M. Vogel, "Assembling a New Hall of Civil Engineering," *Technology and Culture* 6 (1965): 59–73.

8. Alma S. Wittlin, "Hazards of Communication by Exhibits," *Curator* 14 (1971): 145–149.

9. Ibid., p. 146.

10. Ibid., pp. 147–149.

11. Ibid., p. 144.

12. Watson M. Laetsch, "Conservation and Communication: A Tale of Two Cultures," *Southeastern Museum Conference Journal,* March 1979, p. 1.

13. Stephan F. de Borhegyi, "Visual Communication in the Science Museum," *Curator* 6 (1963): 46–47.

14. Guthe, *So You Want a Good Museum,* pp. 27–28.

15. Burcaw, *Introduction to Museum Work,* pp. 121–123.

16. "Sell: Museum Exhibits," *Business Week,* Sept. 30, 1961, pp. 32–33.

17. Howard Learner, "Science Museums: Educational or Promotional?," *Nutrition Act* 6 (1979): 4.

Chapter 16

1. G. Ellis Burcaw, *Introduction to Museum Work* (Nashville: American Association for State and Local History, 1975), p. 129.

2. Ibid.

3. Victor J. Danilov, ed., *Traveling Exhibitions* (Washington, D.C.: Association of Science-Technology Centers, 1978), p. 7.

4. Barbara Tyler and Victoria Dickenson, *A Handbook for the Travelling Exhibitionist* (Ottawa: Canadian Museums Association, 1977), p. 15.

5. Ibid., p. 14.

6. Kerry Marshall, Susan Melin, Janis Eschen, and Alma Sanchez, *Traveling Exhibitions: A Workbook* (Oakland, Calif.: Western Association of Art Museums, 1975), p. 13.

7. Shabtay Levy, in *Traveling Exhibitions,* p. 25.

8. William Brown, in *Traveling Exhibitions,* p. 26.

9. Emily Dyer, in *Traveling Exhibitions,* p. 29.

10. Huntington T. Block, in *Traveling Exhibitions,* p. 31.

11. Antonio Diez, *A Study of an Education Program for S.I.T.E.S.* (Washington, D.C.: Smithsonian Institution Traveling Exhibition Service, 1973), p. 1.

12. Eileen Reynolds, in *Traveling Exhibitions,* pp. 34–35.

13. Watson M. Laetsch, in *Traveling Exhibitions,* p. 39.

14. Lois-ellin Datta, in *Traveling Exhibitions,* pp. 39–40.

15. John Drabik, in *Traveling Exhibitions,* p. 40.

Chapter 17

1. S. Dillon Ripley, introduction in *Museums and Education,* Eric Larrabee, ed. (Washington, D.C.: Smithsonian Institution Press, 1968), p. 1.

2. Frank Oppenheimer, "Schools Are Not for Sightseeing" in *Opportunities for Extending Museum Contributions to Precollege Science Education,* Katherine J. Goldman, ed. (Washington, D.C.: Smithsonian Institution, 1970), p. 8.

3. Ibid.

4. Ibid., p. 9.

5. Ibid., p. 10.

6. Kay Davis and Lee Kimche, *Survey of Education Programs at Science-Technology Centers* (Washington, D.C.: Association of Science-Technology Centers, 1976), p. 10.

7. Ibid., p. 7.

8. Ibid.

9. Ibid., p. 10.

10. C. G. Screven, "The Museum As a Responsive Learning Environment," *Museum News* 47 (1969): 7.

11. Ibid.

12. Davis and Kimche, *Survey of Education Programs at Science-Technology Centers.*

13. *Specials at the Strasenburgh Planetarium,* Rochester Museum and Science Center, Rochester, 1979, pp. 1–2.

14. *Omnimax Theatre/Planetarium,* Imax Systems Corp., Cambridge, Ontario, 1977, p. 1.

15. "Minnesota Museum's Science Film Proves Money-Grabber," *Variety* 182 (1979): 35.

16. Victor J. Danilov, "Libraries at Science and Technology Museums," *Curator* 20 (1977): 98.

17. Ibid., p. 100.

18. Ibid.

19. *Courses, Programs, and Events 1978–79,* Museum of Science, Boston, 1978, p. 4.

20. "Museums-in-the-Mall," *The Cultural Post,* issue 21, Jan.–Feb. (1979), p. 17.

21. Victor J. Danilov, "Museums As Educational Partners," *Childhood Education* 52 (1976): 307.

22. *School Groups 1978–79,* Maryland Science Center, Baltimore, 1978, p. 2.

23. Ibid.

24. Allan D. Griesemer, ed., *Handbook of Programs for Museum Educators* (Lincoln, Neb.: Mountain-Plains Museum Conference, 1977), p. 126.

25. Davis and Kimche, *Survey of Education Programs at Science-Technology Centers,* p. 25.

Chapter 18

1. I. T. Frary, *Museum Membership and Publicity* (Washington, D.C.: American Association of Museums, 1935), p. 6.

2. Richard P. Trenbeth, "Building from Strength Through the Membership Approach," *Museum News* 46 (1967): 25–26.

3. Ibid., p. 25.

Chapter 19

1. George MacBeath and S. James Gooding, eds., *Basic Museum Management* (Ottawa: Canadian Museums Association, 1975; reprint of 1969 ed.), p. 73.

2. Molly Harrison, "Education in Museums" in *The Organization of Museums* (Paris: UNESCO, 1960), p. 81.

3. June Batten Arey, *The Purpose, Financing, and Governance of Museums* (Wayzata, Minn.: Spring Hill Center, 1978), p. 3.

4. Ibid.

5. Ibid.

6. Richard L. Bunning, "A Perspective on the Museum's Role in Community Adult Education," *Curator* 17 (1974): 62.

7. Ibid., pp. 62–63.

Chapter 20

1. Watson M. Laetsch, in *Traveling Exhibitions,* Victor J. Danilov, ed. (Washington, D.C.: Association of Science-Technology Centers, 1978), p. 39.

2. Frank Oppenheimer, interview, Dec. 17, 1981.

3. Joel N. Bloom, in *Measuring the Immeasurable,* Minda Borun, ed. (Washington, D.C.: Association of Science-Technology Centers, 1977), p. i.

4. Ibid.

5. C. G. Screven, "Exhibit Evaluation—A Goal-Referenced Approach," *Curator* 19 (1976): 272–273.

6. Ibid.

7. Ibid.

8. Ibid.

9. Harris H. Shettel, Margaret Butcher, Timothy S. Cotton, Judi Northrup, and Doris Clapp Slongh, *Strategies for Determining Exhibit Effectiveness* (Pittsburgh: American Institutes for Research, 1968), pp. 6–7.

10. Ibid., p. 153.

11. Ibid.

12. Elanna S. Yalow, Randall J. Strossen, Dennis L. Jennings, and Marcia C. Linn, "Improving Museums Through Evaluation," mimeograph, Stanford University, 1979, pp. 3–4.

13. Ibid.

14. Screven, "Exhibit Evaluation—A Goal-Referenced Approach," pp. 274–275.

15. Ibid., pp. 277–284.

16. Arthur W. Melton, "Distribution of Attention in Galleries in a Museum of Science and Industry," *Museum News* 14 (1936): 5–8.

17. Robert B. Bechtel, "Hodometer Research in Museums," *Museum News* 45 (1967): 23.

18. C. G. Screven, "The Museum as a Responsible Learning Environment," *Museum News* 47 (1969): 9.

19. Minda Borun, *Measuring the Immeasurable* (Washington, D.C.: Association of Science-Technology Centers, 1977), p. vii.

20. Ross J. Loomis, "Please! Not Another Visitor Survey," *Museum News* 52 (1973): 21.

21. Terry Dickinson, *Survey of Visitors,* staff memo, Ontario Centre, Toronto, 1975, pp. 1–2.

22. Shabtay Levy, "Visitors and Exhibits at OMSI: A Survey," mimeograph, Oregon Museum of Science and Industry, Portland, 1976, pp. 2–3.

23. A. H. Boerdijk, "Collection and Evaluation of Data Concerning Visitors of the Evoluon," mimeograph, Evoluon, Eindhoven, 1975, pp. 1–8.

24. Ibid., appendixes 1–2.

25. Melton, "Distribution of Attention in Galleries in a Museum of Science and Industry," pp. 7–8.

26. Ibid.

27. Ibid., p. 5.

28. Robert S. Weiss and Serge Boutourline, Jr., "The Communication Value of Exhibits," *Museum News* 42 (1963): 25–26.

29. Laurie P. Eason and Marcia C. Linn, "Evaluation of the Effectiveness of Participatory Exhibits," *Curator* 19 (1976): 46.

30. Ibid., pp. 56–59.

31. Borun, *Measuring the Immeasurable,* p. vi.

32. Ibid., p. xvi.

33. Levy, "Visitors and Exhibits at OMSI: A Survey," p. 3.

Selected Bibliography

Introduction to Science Centers

Alexander, Edward P. *Museums in Motion.* American Association of State and Local History, Nashville, 1979.

Bhathal, R. S., and In, Tian Nguk. *Non-Formal Education in Singapore.* Singapore Science Centre, Singapore, 1980.

Coleman, Laurence Vail. *Company Museums.* The American Association of Museums, Washington, D.C., 1943.

Coulson, Thomas. *The Franklin Institute from 1824 to 1949.* Franklin Institute, Philadelphia, 1949.

Danilov, Victor J. *Starting a Science Center?* Association of Science-Technology Centers, Washington, D.C., 1977.

The Exploratorium: A Synopsis. The Exploratorium, San Francisco, 1980.

Exploring Science. Association of Science-Technology Centers, Washington, D.C., 1980.

Greenaway, F. *A Short History of the Science Museum.* London, 1959.

Greenfield Village and the Henry Ford Museum. Crown Publishers Inc., New York, 1972.

Gwynne, Charles T. *Museums of the New Age.* Association for the Establishment and Maintenance in the City of New York of Museums of the Peaceful Arts, New York, 1927.

Hands-On Museums: Partners in Learning. Educational Facilities Laboratories, New York, 1975.

Hudson, Kenneth. *Museums for the 1980s: A Survey of World Trends.* Holmes and Meier Publishers, New York, 1977.

Kaempffert, Waldemar. *From Cave-Man to Engineer.* Museum of Science and Industry, Chicago, 1933.

Kogan, Herman. *A Continuing Marvel: The Story of the Museum of Science and Industry.* Doubleday & Co., Garden City, N.Y., 1973.

Learner, Howard. *White Paper on Science Museums.* Center for Science in the Public Interest, Washington, D.C., 1979.

National Center for Education Statistics. *1979 Museum Universe Survey.* Institute of Museum Services, Washington, D.C., 1980.

National Research Center of the Arts Inc. *Museums USA.* National Endowment for the Arts, Washington, D.C., 1974.

Oberson, Viola L. *Oregon Museum of Science and Industry.* Metropolitan Printing Co., Portland, 1963.

Olofsson, Ulla Keding, ed. *Museums and Children.* UNESCO, Paris, 1979.

Outhwaite, Leonard. *Museums and the Future.* Institute of Public Administration, New York, 1967.

Panofsky, Walter. *Deutsches Museum—Munich.* Peter-Winkler-Verlag, Munich, no date.

Richards, Charles F. *The Industrial Museum.* Macmillan Co., New York, 1925.

Schwartz, Alvin. *Museum: The Story of America's Treasure Houses.* E. P. Dutton & Co., New York, 1967.

von Miller, Oskar. *Technical Museums as Centers of Popular Education.* Deutsches Museum, Munich, 1929.

Organization and Administration

Access to Cultural Opportunities: Museums and the Handicapped. Association of Science-Technology Centers, Washington, D.C., 1980.

Accounting Principles and Reporting Practices for Certain Nonprofit Organizations. Price Waterhouse & Co., New York, 1978.

Alsford, Denis B. *An Approach to Museum Security.* Canadian Museums Association, Ottawa, 1978.

Anthony, Robert N. *Financial Accounting in Nonbusiness Organizations.* Financial Accounting Standards Board, Stamford, Conn., 1978.

Arey, June Batten. *The Purpose, Financing and Governance of Museums.* Spring Hill Center, Wayzata, Minn., 1978.

Bennett, Paul. *Up Your Accountability.* Taft Products Inc., Washington, D.C., 1973.

Brawne, Michael. *The New Museum.* Frederick A. Praeger Inc., New York, 1968.

Burcaw, G. Ellis. *Introduction to Museum Work.* American Association for State and Local History, Nashville, 1975.

Career Opportunities in Art Museums, Zoos, and Other Interesting Places. U.S. Department of Labor, Employment and Training Administration, Washington, D.C., 1980.

Coleman, Laurence Vail. *Museum Buildings.* American Association of Museums, Washington, D.C., 1980.

Danilov, Victor J., ed. *Museum Accounting Guidelines.* Association of Science-Technology Centers, Washington, D.C., 1976.

Daughtrey, William H., Jr., and Gross, Malvern J., Jr. *Museum Accounting Handbook.* American Association of Museums, Washington, D.C., 1978.

Evelyn, Hugh. *Training of Museum Personnel.* Hugh Evelyn Ltd., London, 1970.

Gross, Malvern J., Jr. *Financial and Accounting Guide for Nonprofit Organization.* Ronald Press Co., New York, 1979.

Guthe, Carl E. *So You Want a Good Museum.* American Association of Museums, Washington, D.C., 1973.

Harrison, Raymond O. *The Technical Requirements of Small Museums.* Canadian Museums Association, Ottawa, 1975.

Hodupp, Shelley. *The Shopper's Guide to Museum Stores.* Universe Books, New York, 1977.

Kotler, Philip. *Marketing for Nonprofit Organizations.* Prentice-Hall Inc., Englewood Cliffs, N.J., 1975.

MacBeath, George, and Gooding, S. James, eds. *Basic Museum Management.* Canadian Museums Association, Ottawa, 1975.

Miller, Ronald L. *Personnel Policies for Museums: A Handbook for Management.* American Association of Museums, Washington, D.C., 1980.

Museum Management Tools. Management Analysis Center Inc., Washington, D.C., 1978.

Naumer, Helmuth J. *Of Mutual Respect and Other Things.* American Association of Museums, Washington, D.C., 1977.

The Organization of Museums. UNESCO, Paris, 1967.

Perry, Kenneth D., ed. *The Museum Forms Book.* Texas Association of Museums, Austin, 1980.

Seymour, Harold J. *Designs for Fund-Raising.* McGraw-Hill Book Co., New York, 1966.

Swinney, H. J., ed. *Professional Standards for Museum Accreditation.* American Association of Museums, Washington, D.C., 1978.

Thomson, Garry. *The Museum Environment.* Butterworth & Co., London, 1978.

Thoughts on Museum Conservation. Texas Historical Commission, Austin, 1976.

Thoughts on Museum Planning. Texas Historical Commission, Austin, 1976.

Tillotson, Robert G., and Menkes, Diana D., eds. *Museum Security.* International Council of Museums, Paris, 1977.

Ullberg, Alan D., and Ullberg, Patricia. *Museum Trusteeship.* American Association of Museums, Washington, D.C. 1981.

Workshop on the Establishment of Science Museums in Asian Countries: Training and Exchange. UNESCO, National Council of Science Museums (India), ICOM Regional Agency in Asia, and Indian National Committee for ICOM, Bangalore, India, 1981.

Exhibits and Programs

Borun, Minda. *Measuring the Immeasurable.* Association of Science-Technology Centers, Washington, D.C., 1977.

Borun, Minda, and Miller, Maryanne. *What's in a Name?* Franklin Institute, Philadelphia, 1980.

Bruman, Raymond. *Exploratorium Cookbook.* The Exploratorium, San Francisco, 1975.

Casterline, Gail Farr. *Archives & Manuscripts: Exhibits.* The Society of American Archivists, Chicago, 1980.

Chenhall, Robert G. *Museum Cataloging in the Computer Age.* American Association for State and Local History, Nashville, 1975.

Clasen, Wolfgang. *Expositions, Exhibits, Industrial and Trade Fairs.* Frederick A. Praeger Inc., New York, 1968.

Cole, K. D. *Vision: In the Eye of the Beholder.* The Exploratorium, San Francisco, 1978.

Danilov, Victor J. *Survey of Libraries.* Association of Science-Technology Centers, Washington, D.C., 1975.

Danilov, Victor J., ed. *Traveling Exhibitions.* Association of Science-Technology Centers, Washington, D.C., 1978.

Davis, Kay, and Kimche, Lee. *Survey of Education Programs at Science-Technology Centers.* Association of Science-Technology Centers, Washington, D.C., 1976.

Directory of Exhibits at Science & Technology Centers. Association of Science-Technology Centers, Washington, D.C., 1976.

Dudley, Dorothy, H., and Bezold, Irma. *Museum Registration Methods.* American Association of Museums, Washington, D.C., 1958.

Dudley, Dorothy H., Wilkinson, Irma Bezold, et al. *Museum Registration Methods.* American Association of Museums, Washington, D.C., 1979.

Fox, Martin, and Carpenter, Edward K. *1980–81 Edition: The Best in Exhibition Design.* RC Publications Inc., Washington, D.C., 1980.

Frary, I. T. *Museum Membership and Publicity.* American Association of Museums, Washington, D.C., 1935.

Griesemer, Allan D., ed. *Handbook of Programs for Museum Educators.* Mountain-Plains Museum Conference, Lincoln, Neb., 1977.

Grinell, Sheila, ed. *A Stage for Science: Dramatic Techniques at Science-Technology Centers.* Association of Science-Technology Centers, Washington, D.C., 1979.

Hands On: Setting Up a Discovery Room in Your Museum or School. Royal Ontario Museum, Toronto, 1979.

Hipschman, Ron. *Exploratorium Cookbook II.* The Exploratorium, San Francisco, 1980.

Humphrey, Thomas, and Tanner, Marcia. *A Guide to the Exploratorium Exhibits I.* The Exploratorium, San Francisco, 1977.

Larrabee, Eric, ed. *Museums and Education.* Smithsonian Institution Press, Washington, 1968.

Lewis, Ralph H. *Manual for Museums.* United States National Park Service, Washington, D.C., 1976.

Mankind Discovering: Volume II: Evaluations—The Basis for Planning. Royal Ontario Museum, Toronto, 1979.

Marshall, Kerry, Melin, Susan, Eschen, Janis, and Sanchez, Alma. *Traveling Exhibitions: A Workbook.* Western Association of Art Museums, Oakland, 1975.

Museums, Imagination and Education. UNESCO, Paris, 1973.

Neal, Arminta. *Exhibits for the Small Museum.* American Association for State and Local History, Nashville, 1978.

Rebetez, Pierre. *How to Visit a Museum.* Council for Cultural Cooperation of the Council of Europe, Strasbourg, 1970.

Rhees, David. *Creating Museum Exhibits about Computers: A Guide to Resources.* Association of Science-Technology Centers, Washington, D.C., 1981.

Stolow, Nathan. *Procedures and Conservation Standards for Museum Collections in Transit and on Exhibition.* UNESCO, Paris, 1981.

Taylor, David, and Rhees, David. *Survey of Computer Use in Science-Technology Museums.* Association of Science-Technology Centers, Washington, D.C., 1981.

Temporary and Travelling Exhibitions. UNESCO, Paris, 1963.

Trade Shows and Exhibits. Association of National Advertisers Inc., New York, 1968.

Tyler, Barbara and Dickenson, Victoria. *A Handbook for the Travelling Exhibitionist.* Canadian Museums Association, Ottawa, 1977.

Warren, Jefferson T. *Exhibit Methods.* Sterling Publishing Co., New York, 1972.

Weiss, Robert S., and Boutourline, Serge, Jr. *Fairs, Pavilions, Exhibits, and Their Audiences.* Robert S. Weiss, New York, 1962.

Witteborg, Lothar P. *Good Show! A Practical Guide for Temporary Exhibitions.* Smithsonian Institution, Washington, D.C., 1981.

Index